Intruder Alarms

Intruder Alarms

Third edition

Gerard Honey

AMSTERDAM • BOSTON • HEIDELBERG • LONDON • NEW YORK • OXFORD
PARIS • SAN DIEGO • SAN FRANCISCO • SINGAPORE • SYDNEY • TOKYO

Newnes is an imprint of Elsevier

Newnes is an imprint of Elsevier
Linacre House, Jordan Hill, Oxford OX2 8DP, UK
30 Corporate Drive, Suite 400, Burlington, MA 01803, USA

First edition 1997
Reprinted 1999, 2000
Second edition 2003
Reprinted 2005, 2006
Third edition 2007

British Library Cataloguing in Publication Data
A catalogue record for this book is available from the British Library

Library of Congress Cataloging-in-Publication Data
A catalog record for this book is available from the Library of Congress

ISBN–13: 978-0-7506-8167-4
ISBN–10: 0-7506-8167-5

For information on all Newnes publications
visit our website at www.newnespress.com

Typeset by Charon Tec Ltd (A Macmillan Company), Chennai, India
www.charontec.com
Printed and bound in *Great Britain*

07 08 09 10 11   10 9 8 7 6 5 4 3

Contents

Foreword

Over the last decade, considerable progress has been made in the development of educational systems, qualification frameworks and a sector qualifications strategy, all of which are designed to support the rapidly changing Security Business Sector that has grown both in size and significance, as organisations and individuals strive to protect their people, property and other assets.

The City and Guilds 1852 examination, supported by Skills for Security, the Skills and Standard Setting Body for the sector, is an example of a qualification that has been specifically designed to help the security systems industry develop the pool of technicians required, if it is to meet the demands of its customers and respond to technological and operational changes in the marketplace.

Skills for Security is, therefore, delighted to endorse this book as a supporting text for the City and Guilds 1852 syllabus. In addition, we believe that the book is also a useful companion to the 1882 National Vocational Qualification scheme; Providing Security Emergency and Alarm Systems at Levels 2 and 3.

We are certain that you will find this book of considerable assistance to your understanding of the security systems industry and alarm technology in general, and wish those of you who proceed to take qualifications every success in your endeavours.

Stefan Hay FSyI
Director of Strategic Development
Skills for Security

Preface

This book can be seen to be the definitive guide to intruder alarm systems since it extends its scope to consider all of the controlling practices together with the various regulations and documents that apply in this industry. In addition it has been written to follow the City and Guilds syllabus 1852 Level 3 Certificate of Knowledge of Security and Emergency Alarm Systems module 001 Intruder Alarms. Therefore whilst it contains the essential content of this governing systems course it can also be used by all personnel with an interest in this sector on an international basis.

The chapters of the book have been introduced as separate units which form the framework of the intruder alarm industry. They will enable the reader to:

- apply new security systems specification to physical installation and customer requirements;
- establish the operational feasibility of planned security systems;
- install a complete security system;
- install and test security system equipment which uses data transmission protocols;
- commission and handover operational security systems;
- audit installed security systems;
- diagnose and rectify faults in security systems;
- contribute to on-site health and safety;
- communicate effectively with others.

Acknowledgements

The author would like to thank all those who provided information for this book, and in particular:

ACT Meters
Castle Care Tech Ltd
Cooper Menvier Security (Scantronic)
FM Electronics
JWM Control Systems
National Security Inspectorate (NSI)
Pyronix
Security Systems and Alarms Inspection Board (SSAIB)
Skills for Security. Formerly Security Industry Training
   Organisation (SITO)
Tavcom Training Ltd

1 Intruder alarm systems

This chapter concentrates on the fact that intruder alarms form only one part of a much greater security industry. The need for confidentiality in the industry cannot be overemphasized. Technicians need to be aware of the risk to themselves, their employers and of course their customers, and it is important to understand each person's role in the intruder alarm industry sector.

It is certainly true to say that the last few years have seen many changes in the security industry, and advanced technologies combined with stricter standards and working practices will continue this trend. Technologies in the security industry will become more complementary, and the engineer involved with alarms must understand how this will affect him or her and what can be expected in the long term.

The industry for installers has changed dramatically because of the introduction of the European Standards and the ongoing influence of the Association of Chief Police Officers (ACPO) Security Systems Policy. The year 2000 version of the policy, which is concerned with all security systems that may generate an 'electronic call for help', brought radical changes to previous versions. Indeed the current edition of the policy has regularly been under scrutiny and was last critically reviewed in October 2005. The ideal is to reduce false alarm activations that are passed to the police but with the emphasis being placed on the industry itself to provide solutions. In keeping with the aims of the policy the security systems sector is expected to create standards and codes of practice to achieve results by means of the current technologies available. Within this DD 243: 2004 has been introduced as a code of practice to give guidance and advice on the design, installation and configuration of intruder alarm systems which signal confirmed alarm signals to an alarm receiving centre (ARC). This new policy places great emphasis on the use of such confirmed alarm technology in seeking the verification of alarms on monitored systems.

We have also seen the creation of the National Security Inspectorate (NSI) and many changes within the schemes that provide the inspection services for companies installing electronic security devices. In addition the security industry has shown great progress in the development of educational systems and qualification frameworks in tandem with the changes in the industry. How can we expect the installer's role to develop in the next decade? The verification of alarms and integration with closed-circuit television (CCTV) and access control is certainly apparent as is also a higher emphasis on the external protection of sites, whereas the past has seen a greater use of internal protection systems. More capital is expected to be spent on making sites more secure, and training will become more important as customers look for management systems. These buyers are already seeking more computer-based systems, and so this is a natural progression.

Certainly there is an increased demand for perimeter intruder detection systems (PIDSs), which offer an earlier warning of potential crime than purely internal detection. Integration will also play a major role in the security industry, and stand-alone technologies will become outdated, certainly in the commercial and industrial market. It will then progress into the residential domestic arena. We will, additionally, witness a growing use of personal computers (PCs) in the installation business: initially the security industry used DOS basics but Windows-based systems are becoming more common. Indeed, as more computer-based systems come on line the days of traditional programming are now numbered, and software development will be achieved remotely by central offices using modems for uploading and downloading programs.

Standards, policies and regulations will all change radically as Euro Norms (EN standards) are adopted, but as policies and regulations are enhanced it will be necessary to dedicate more time to administrative work. As the equipment becomes increasingly technical there must be a corresponding increase in training, particularly with more PC-based work. System types must surely become large, and with these being PC derived the complexity will escalate. The technology we expect to see developing through the coming years will certainly be the observation system integrated with intruder functions. The systems will become increasingly refined as prices of surveillance components fall, and the development of new technologies will enable high-resolution kits to be more affordable. Certainly there will be a huge amount of domestic premises boasting CCTV-type systems in the not-too-distant future. Intruder alarm systems with video access control and similar technologies are definitely something that architects and planners must now consider, and the provision of this security within new buildings will become a 'must'.

Intruder alarm systems will become a part of building management systems or intelligent buildings, and the emphasis will not only come from the security industry but from architects, planners and developers. The security installer must ensure that he or she fits into this scheme, and the developments, or face losing the market to electricians and builders.

Changes in technology will lead to changes in the knowledge that is required by the installer, and it is important to predict the route that security installations will take in the coming years. Eventually the communication side of the industry will be changed dramatically and there will be a great deal of emphasis on the uploading and downloading of data. Those installing alarms must diversify. And as far as equipment is concerned, its reliability must improve. So there it remains: we must expect integration and technological advancement. The market for intruder alarms will continue to grow but as an integrated technique with true compatibility of system hardware.

1.1 A complementary part of the security industry

Security systems range through intruder alarms to access control and CCTV but with an interest also in lighting methods to embrace automatic detection, occupancy and emergency. There is also a small level of integration with call systems, including nurse call systems plus manual and automatic fire detection and associated alarms.

No matter how we look at intruder alarms we can certainly see how they become involved with access control and CCTV as Figure 1.1 shows.

Figure 1.1   *Intruder/access control/CCTV*

Different lighting forms are often integrated with the three principal systems, be it occupancy, automatic or emergency, and fire detection then becomes a related subject.

In practice it will be found that an installation company performing the fitting of intruder alarms will recognize the importance of being aware of access control and CCTV, and will be familiar to some extent with call and fire alarms plus lighting methods.

In reality, outputs may be taken from a system to activate another. Remote signalling equipment provides different channels for the various functions transmitted to alarm receiving centres (ARCs), and verification of intruder alarms can be achieved by the use of lighting and CCTV. In cases where no specific outputs exist, electromagnetic relays can be energized to open or close circuit loops. These relays are discussed later. The integration of systems is wide ranging, and although the application of intruder and fire follow somewhat different practices, an understanding of one helps to bring an understanding of the other. The description of line monitoring and switching techniques in either type facilitate an understanding of why these practices are adopted for alarm systems in buildings.

The security industry has and will continue to witness unprecedented change that has many repercussions for the installer and central station or alarm receiving centre which will monitor signals, not only from intruder alarms but from a great many other sources also.

The intruder alarm installer must recognize how his or her industry complements others and realize the key purpose of security: to protect life, premises and property. It also leads us to the manned section of the industry (static and patrol guarding), comprising:

- aviation security;
- property under guard;
- retail guarding;
- store detection;
- close protection;
- dog handling;
- physical protection.

The manned section, although not needing to understand the technological detail, should be aware of the other security areas and the roles they play. Those installing intruder alarms must equally understand the role of the manned sector and how it will often be connected to the alarm receiving centre to follow up intrusion outputs.

Crime is a problem we certainly live with. It affects us in different ways so it cannot be easily priced or defined. In reality we hear different figures and statistics from different sources but what we do know is that the problem will never go away. We also know that a lot of crime is fuelled by increasing drug use which does account for high levels of shop theft, as reported by the British Retail Consortium.

Engineering and technical staff alongside the manned section and together with the central station and a good input of police response can address our present-day problems and suggest plans for the future of our industry. Within this framework it is vital that all those involved with intruder alarm technologies recognize the roles that they must play.

1.2 European Standards Euro Norms (EN). EN50131

In practice the intruder alarm security industry is governed by many standards, codes of practice and various industry documents. However it is the European Standards in the form of EN50131 that are the most important documents so it is to these that we must first turn our attention.

As from 1 October 2005 all intruder alarm installations in the UK became governed by new European Standards or Euro Norms (EN) which replaced the traditional British Standards. In practice ENs exist so as to remove barriers to trade by providing a common set of standards for Europe as a whole.

The withdrawal of the traditional and time honoured BS 4737 in favour of the EN50131 suite of standards had a major impact throughout the intruder alarms industry. EN50131-1 became known as the pan-European standard for intruder and hold up alarm systems with Part 1 being the general system requirements. All intruder security installation inspections became subject to the new EN50131-1: 2004 standard with PD6662: 2004 being a guide that would determine how the scheme would be implemented until the full suite of standards were to become finalized. This was necessary because at the time of the introduction date not all parts of EN50131 were available. Therefore, in the UK, security installations were indicated as being in compliance with PD6662: 2004 rather than to EN50131-1.

It is important to remember that from October 2005 all personnel involved with intruder alarms became committed to observe the ENs when risk assessing, specifying, installing and maintaining a new system. All existing British standards which conflicted with the EN standards had to be withdrawn; however, because of the long history of certain BS standards, the influence that they had held was still felt for some time and this did reflect in the content of PD6662.

EN50131

The suite of standards, which are covered by EN50131, are the most important documents to all those involved with the intruder alarm systems. The major difference between the new European Standards and the extinct BS 4737 was the introduction of grading for alarms. This is because the ENs were structured standards that are risk based. Therefore the particular security grades for the installation were to be in relation to the risk faced and had to be determined by the installer. This meant that the alarm engineer had to be trained to assess risks in a much more formal capacity than that demanded of the original generation British Standards.

However alongside EN50131 two other documents became of vital importance to the installer and these are namely PD6662: 2004 and DD 243: 2004. Both were published for consultation in the late part of 2004 and indicate how the Inspectorates and the Police will interpret and expect the requirements of the ENs to be carried out.

PD6662 is essentially a wrapper type document specifying documents for the EN50131 scheme in the UK and with some requirements brought over from the traditional but now withdrawn BS 4737.

DD 243 relates to alarm confirmation practices and the possible procedures for transmitting confirmed alarms through to an Alarm Receiving Centre.

PD6662

The actual process of introducing European Standards for the security alarm industry was started in the 1990s with the idea of bringing together the best parts of each country's national standards so as to create a Euro Norm. The standard in which the intruder alarm industry had an interest came to be known as EN50131 with part 1 (EN50131-1) covering the general System Requirements. However in order to replace the existing national standards meant that a suite of EN50131 standards was needed. This proved somewhat difficult to instigate and the process of generating the EN and gaining approval for it throughout Europe led to a quick fragmentation of the full suite of documents. In the early stages not all documents were available and others were ambiguous. The result was the introduction of a dual-running period with EN50131 and BS 4737 running alongside each other for an interim period. Alongside this the UK industry created a guide in the form of PD6662: Scheme for the application of European Standards for intruder and hold-up alarm systems. PD6662 was used in order to determine how the EN scheme could be used until all documents that were to form the full suite of EN50131 were available. This also meant that in the early days UK security installations would be indicated as compliant with PD6662 rather than EN50131.

Therefore we can say that for the intruder alarm industry PD6662 was produced as an initial response to the large gaps in the EN Standards/EN50131 that existed on their introduction into the UK.

PD6662 can be said to be the 'key' document to implementing the EN Standards in the UK.

With all new installations now specified against PD6662:2004 this became known as the largest transformation of the UK intruder alarms industry in some 30 years.

The PD is published by the British Standards Institution (BSI) and as we now recognize describes the UK scheme for the implementation of the European Standards (EN50131-1 etc.) for intruder and hold up alarm systems.

Specifiers who may be more aware of EN50131 than PD6662 should note that the PD scheme calls up a range of standards including some BS 4737 component standards. Therefore tender documents tend to refer to PD6662 rather than EN50131.

The lead standard in the EN50131 series called up by PD6662 is the 2004 draft edition of Part 1 of EN50131, referred to as prEN50131-1: 2004. It is this which requires PD6662 alarms to be given a security grade from 1 to 4 with Grade 1 being low risk and Grade 4 being high risk.

From this it is important to remember that the pr prefix identifies a draft EN document.

Technical Specifications (TS)

These were drawn up as the starting process of the conversion to an EN standard and represented individual product standards for particular components and equipment. Identified by the prefix TS they did not have the same standing as an EN. In October 2006 PD6662 was called up for modification to invoke a TS as though it was an EN and any equipment sold would need to meet the requirements of the TS. For the installer it gave a product designed to the appropriate TS that would apply when the EN was ratified without a need to then install new EN compliant equipment.

BSIA Form 185 provided recommendations to manufacturers for the interpretation of the all important detector standards DD TS 50131-2-2: 2004 covering passive infrared detectors and DD TS 50131-2-4: 2004 covering combined passive infrared and microwave detectors. It was also found that manufacturers would claim their products as compliant to the relevant TS as opposed to claiming PD6662 or EN50131 compliance. This gave an advantage in not having to change a detector for instance when the UK formally adopted the individual TS individual product standards into PD6662.

Terminology

A certain amount of changes to the terminology used in the intruder industry occurred when the EN standards replaced the BS standards. In this book various terms will be used simply because the industry still uses various terms. The most encountered are:

Zones. These originally meant a detection circuit. Now are often called inputs. Zones in PD6662 actually refer to what we call areas and zones under the ENs are known as assessed areas.

Resets. Engineer resets are often labelled as restores.

PA. Personal attack. Now called hold up alarm triggering device or HU.

STU. Subscribers terminal unit and **Digi** (an abbreviation for digital communicator) all tend to be called alarm transmission equipment (ATE).

EN50131 structure

The significant difference between the withdrawn BS 4737 and its replacement EN50131 is that the latter is risk based with a format structured as follows:

Assessment of risk

Technical survey

System design

Installation of a system in accordance with an agreed specification

Installation of equipment to the manufacturers' recommendations

With this structure in mind, working with the EN Standards, means that the following form of approach is needed:

- Risk assessment in relation to the area/premises to be protected
- Technical requirements survey in a formal manner
- System design – System design proposal specification
- Installation of a system in accordance with an agreed specification using appropriate equipment and components
- The installation of equipment in accordance with the manufacturers recommendations and limitations on use

Therefore the steps required, in order to work to the ENs in their most simple terms are as follows, with the procedure used to design and record a system based on risk:

DESIGN A SYSTEM BASED ON RISK

Carry out a risk assessment
Take into account
Risk level of the premises
Building location and design
Contents value
Minimum supervision levels required
Typical knowledge intruders are expected to have
Technical requirements

Determine a risk grade (1, 2, 3 or 4)

Detail and record the findings of the risk assessment

Produce a system design – System Design Proposal specification

Install manufacturers' compliant products to the risk grade (equipment and components)

Carrying out the risk assessment, detailing and recording the findings and producing the system design is an involved process. The way this can be organized is covered in Section 1.3 of this book together with information as to how the system grading is applied in the industry.

It will be noticed that the installer must fit equipment and system components that are manufactured as being compliant with the specific grade for the system that is to be installed. This is a definite EN requirement.

The grading of the system also determines the performance of many things. Included in this are Environmental Classes, Authorization, Access levels, Operations, Processing, Detection, Notification, Power supply, Tamper security, Event Recording and Monitoring of interconnections.

In order for an installation to be actually defined as a system under the ENs it must include the following components:

● Control and indicating equipment
● One or more detectors

- One or more warning devices and or an alarm transmission system
- One or more power supply units combined with other system components or housed separately
- Interconnections by cable, wireless or any other method providing it gives the required monitoring for the appropriate grade
- System components classified according to their environmental capability and graded relative to their performance
- Components of other applications may be integrated with a system providing the performance of the system components are not adversely influenced

Throughout the various Sections of this book information will be given as to how the grading is applied for the different system parts and for the system roles. However it will be seen that all of the actual system components must also be classified according to their environmental capability and given an *Environmental Class*. This is in addition to the products being risk graded relative to their performance and proved compliant by the manufacturer to such a grading.

Other general requirements of any system involve *Access Levels* and *Event Recording*. These are all applied as follows and in a particular way:

Environmental Classes

There is a need to select a product classified according to its Environmental Class numbered I, II, III or IV listed in a progressively more severe or harsh environment order. All components must be classified to meet, or better, the environmental criteria for the location in which it is to be sited.

- **Class I.** Indoor controlled environments. For general use where the temperature is maintained.
- **Class II.** Indoor environments where the temperature may vary more than with Class 1. Examples are unheated storage areas and those places in which the heating is intermittent.
- **Class III.** Outdoors sheltered. The units are not to be fully exposed to the weather.
- **Class IV.** General outdoor use with exposure to the weather.

Access Levels

Four access levels are defined and determine the people who have access to the system. Each group will need to access different functions of the system and certain functions may be accessed by more than one level.

Access Level 2 is the normal and familiar user codes and provides the time honoured normal master and manager levels.

Access Level 3 is the installer/service code. In some instances a customer code (Level 2) is needed to authorize the logging on of a Level 3 code.

Access Level 4 is intended to enable manufacturers to make software changes and is not intended to change or programme system parameters.

- **Access Level 1.** Any person
- **Access Level 2.** Users and operators of the intruder alarm system
- **Access Level 3.** Any representative of the installation or servicing company
- **Access Level 4.** Any person representing the manufacturer

Event Recording

The number of events that are to be recorded in the system control equipment log depends on the system grade. If power is lost the memory is to be capable of retaining the information for more than 30 days.

- **Grade 1.** Optional
- **Grade 2.** 250 or more events
- **Grade 3.** 500 or more events
- **Grade 4.** 1000 or more events

General Requirements

There are two fundamental documents that all of those involved with intruder alarms need to be particularly familiar with, namely EN50131-1 (General Requirements) and EN50131-7 (Application Guidelines).

EN50131-1: 2004 is the General Requirements document and covers all of the system requirements and details the functions and how the system is to operate. It details what the system is to comprise of and how the system should operate.

It is of 12 main sections covering:

- System functions
- System components
- Security grading
- Environmental classification
- Functional requirements
- Power supply
- Operational reliability
- Functional reliability
- Environmental requirements
- Electrical safety
- Documentation
- Marking and identification

EN50131-7 is the Application Guidelines and provides advice in regard to the system design. In addition it refers to the operation and the maintenance of alarm systems.

It is made up of 8 main sections covering:

- System grading
- Environmental classification
- System design, installation and planning
- System installation
- Inspection test, commissioning and acceptance
- Documentation and records
- Operation
- Maintenance and repair

From our observations on the content of the EN50131 suite of standards it illustrates how involved the intruder alarm engineer must become so as to make an assessment and ultimately compile a System Design Proposal.

System Design Proposal

This is to follow the Application Guidelines so must consist of:

- Grade of the system and of any subsystem
- The Environmental Class of all components
- The schedule of the equipment. This depends largely on the system size and complexity. Drawing should be used to illustrate the location, building and detection coverage
- System configuration. To detail the programming and circuit types together with entry routes etc.

Documentation. Marking and identification

The documents that are produced for the intruder alarm system are to be clear and concise.

They must also be unambiguous and complete with the correct level of information needed to install, operate and maintain any system.

The components for installation must also be identified and contain as a minimum:

- Name of manufacturer or supplier
- Description
- Standard to which the products claim compliance
- Name or mark of the certification body if applicable
- Environmental Class

The introduction of the EN Standards brought a greater level of consistency to the specifications of the products to be installed for the different security grades and established an industry level for the services demanded by the end user.

The preventative maintenance requirements were set so as to provide for a greater use of remote system checks and the frequencies that apply are contained in Chapter 9 of this book. Systems were also pioneered to flag up problems in order to advise when its performance fell below a pre-determined operational threshold with diagnostic information becoming the norm.

Conclusions

The introduction of EN Standards brought many changes to the intruder alarm industry.

It is to be said that they are risk based and this places greater emphasis on the installation company to carry out a risk assessment.

The components must then be selected to suit the security grade that has been established. All intruder alarm components and equipment are to be classified to meet an appropriate grade by the manufacturer or supplier, so these must be selected as appropriate for the proposed system by the installer. The installation company is obliged to detail the findings from the risk assessment and produce the system design proposal. All of this information is to be recorded. Equally the installer must understand all of the other factors that govern the installation of products for the selected security grade. These factors appear in the various sections of this book.

The standards used throughout the industry in general are contained in Section 10.4.

The chart depicted as Table 1.1 can be used as an overview to see how the introduction of EN50131 first became operational in the UK. This chart has been further reproduced in Section 10.4 in order to be listed alongside the other industry control documents.

1.3   Security grading

As we now know the withdrawal of BS 4737 and the introduction of the EN50131 suite of standards had a major impact throughout the intruder alarms industry. This was because the ENs were structured standards that are risk based. It introduced a rather different approach in the way that the UK intruder alarm industry was now to be governed by the controlling documents.

It meant that the particular security grades for the installation were to be in relation to the risk faced and had to be determined by the installer.

In effect the alarm engineer had to be trained to assess risks in a much more formal capacity than that demanded of the original generation British Standards.

Table 1.1   *Standards and Documents for "EN" Intruder Alarm Systems (October 2006 – As of January 2006)*

Document		Number	Called up by PD 6662	Description	BSIA guidelines Form No.
UK Prefix*	Doc type*				
PD		6662: 2004		Wrapper document specifying documents for the EN50131 scheme in the UK and with some requirements brought over from BS 4737	Form 171 Issue 2
DD CLC/	prEN	50131-1: 2004	Yes	*System Requirements*	
DD CLC/	TS	50131-2-2: 2004	Yes[a]	PIR	Form 185
DD CLC/	TS	50131-2-3: 2004	Yes[a]	Microwave	Form 185
DD CLC/	TS	50131-2-4: 2004	Yes[a]	PIR/Microwave	Form 185
DD CLC/	TS	50131-2-5: 2004	Yes[a]	PIR/Ultrasonic	Form 185
DD CLC/	TS	50131-2-6: 2004	Yes[a]	Opening door Contacts (magnetic)	Form 184
DD CLC/	TS	50131-3: 2003	Yes[a]	*Control & Indicating Equipment (CIE) (i.e. control panels)*	Form 179
DD CLC/	TS	50131-5-3: 2005	Yes[b]	Wirefree Interconnections (RF) standard	
BS	EN	50131-6: 1997/8	Yes	*Power Supplies*	Form 180
DD CLC/	EN	50131-7: 2003	Yes	*Application Guidelines (i.e. for installers)*	
BS	TS	50136-1-1: 1998	Yes	Alarm Transmission Systems (ATS) (i.e. comms to ARC) General	
BS	EN	50136-1-2: 1998	Yes	Using dedicated paths	
BS	EN	50136-1-3: 1998	Yes	Using PSTN	

BS EN	50136-1-4: 1998	Yes	Using voice communicators	
BS EN	50136-2-1: 1998	Yes	*Alarm Transmission Equipment* (ATE) General	
BS EN	50136-2-2: 1998	Yes	Using dedicated paths	
BS EN	50136-2-3: 1998	Yes	Using PSTN	
BS EN	50136-2-4: 1998	Yes	Using voice communicators	
DD	243: 2004	Yes	*Alarm Confirmation*	
BS	4737-3-X (excl. –3–30)[c]	Yes	Specifications for *Components* (i.e. parts without an EN)	
		No	"Manuf. Agreement *Additional Comms Req.* between ATE & CIE"	Form 175
		No[d]	"ACPO Requirements For The Reduction of *False Hold-Up Alarms*"	Form 176
		No	Commissioning and Preventative Maintenance (*remote servicing*)	Form 177

* BS = British Standard, CLC = CENELEC, DD = Draft for Development, EN = European Standard, PD = Published Document, pr = Draft, TS = Technical Specification, [a] Included as of 1st October 2006, [b] Included as of January 2006, [c] Also excluding 4737-3-4 and 4737-3-7 as of 1st October 2006, [d] Included in ACPO Security Systems Policy

Table 1.2 *Security grading*

GRADE 1. Low risk
Intruders are expected to have little knowledge of intruder alarm systems and be restricted to a limited range of easily available tools
Essentially DIY market

GRADE 2. Low to medium risk
Intruders are expected to have a limited knowledge of intruder alarm systems and use a general range of tools and portable instruments
Residential premises. Professional installations
Note: 2X is a grade identified in the UK industry which allows for non-communicating Grade 2 systems. All other requirements of Grade 2 remain.

GRADE 3. Medium to high risk
Intruders are expected to be conversant with intruder alarm systems and have a comprehensive range of tools and portable electronic equipment
Commercial premises. Professional installations

GRADE 4. High risk
To be used when security takes precedence over all other factors. Intruders are expected to have the ability or resources to plan an intrusion in detail and have a full range of equipment including means of substitution of vital components in the intruder alarm systems.
Equipment for this grade is specialised
Bespoke high security systems

The security grading is to Table 1.2:

There is no particular method for conducting a risk assessment but it is necessary to ensure that all risks are considered and a system be designed to protect that risk.

However a full description of risk assessment guidance is given in the UK prefixed document:

DD CLC/TS 50131-7: 2003 Intruder Alarm Systems – Application Guidelines.

It is also worth noting that Clause 6.5 of the National Foreword of this document states that all persons responsible for the risk assessment and the design, installation, maintenance and repair of intruder alarm systems should hold appropriate qualifications. Although there may be no defined agreed qualifications structure it does quite clearly mean that the competence of all interested parties must be capable of being demonstrated.

To establish the level of risk and subsequent security grade for the premises to be protected the installation company needs to consider different criteria involving:

● Location
● Contents

- Building design
- Minimum supervision levels required

Risk assessment

A risk assessment must be carried out for every installation under the EN requirements.

A number of stages need to be considered in this risk assessment since once the grade of system has been established it determines what the intruder alarm system will actually comprise of.

Evaluating the need must be carried out in a structured format.

A simple form of industry technique in order to do this is to use the **ABC** of risk where:

A = Area
B = Boundaries/buildings
C = Contents

As a means of gathering information a pre-survey needs to be carried out by talking to the client. This gives the client an opportunity to outline the perceived problems. Following the pre-survey a fuller report can then be made.

PRE-SURVEY

Area

Rural, suburban, inner city or commercial
Does the premises have easy access to the main national roads and motorways
How easy do criminals find ease of movement and the road/rail infrastructure
Do criminals target the local commercial sites
Is the crime rate high or low
The insurance industry will set its premiums in relation to the history of the area

Boundary/buildings

Are any security systems already in force as a boundary is not always physical
Are all physical restraints in good repair
Can the boundary be overcome by scaling and is material available to use to such effect
Do trees and other building parts allow climbing to be employed
Is adequate external lighting available and is it efficient
Are doors and windows in good repair
Are adequate locks fitted
Are external walls solid

Contents

What are they
Where are they
Are they easily transported
Is there a security guard
Does the security guard have unrestricted access to all areas
Is the security 24 hours
Is any access control used to limit movement to areas with particular contents
and is there any monitoring by such techniques as CCTV

The simplest method of calculating the actual risk is to use a 'Surveyors risk analysis checklist'. To ensure that the correct information is available be methodical and try to use the same system. Always check the basics.

Every risk analysis will have its own peculiarities therefore a checklist is required as a starting point towards a more detailed and specific examination.

A number from 1 to 4 should be put against each section of the risk with:

1 = low and 4 = high.

Checklist examples follow for how the three ABC risk categories can be formed.

Using a format based on such logic will enable the surveyor to then comprise their own in-house forms and charts.

It will be noted in the checklist example for 'Area' that because the lighting is good the risk number is low because of the value of the lighting to help the security aspects. However because the road infrastructures are good in this instance it means transportation of goods from the site is easy so the security risk becomes higher in relation.

Further checklists can be formulated for 'Boundary/buildings' and 'Contents' with comments included as appropriate.

Area

	Risk grade
Rural urban inner city ✓ other	3
Inner city with high density population	
Road network to/from premises	4
Easy access	
Lighting in general location	1
Good levels of high quality light available	
Similar premises in location	2
Other similar engineering fabrication plants	
Previous intrusion rates	2
Low incidence	

Rail/road infrastructure	4
Advance networks and intersections	
Vandalism history	3
Slightly higher than normal need for Personal attack devices	

Area overview. Comments

An industrial area on fringes of an inner city. High levels of activity throughout the day. Passing traffic outside of normal working hours. An area that is often frequented by youths on the evening.
Personal attack (PA) devices are required.

Boundary/buildings

Risk grade

Perimeter
Good/poor repair	2
Adequacy	3

Construction
Common perimeter fences
Rivers adjacent to buildings
Vulnerability of telephone lines on
building/external perimeter

Lighting
Perimeter lit at night
Lighting beyond perimeter
Protection against vandalism
Back-up supply

Position and occupancy
Does shrubbery obscure building
Do adjacent buildings provide an entry point
Is the building shared occupancy
Do rooms/areas facilitate change of use

Physical condition of structures
Height of building
Doors/windows/walls/roof/frames
Skylights/locks/service chutes
Flat roofs

Scaling aids
Trees
Drainpipes
Scaffolding
Chemical drums
Railings
Stored plant to aid scaling
Other climbing points

Boundary access
Manned security
Vehicle access control
Car park for other users
CCTV in use
Mobile security site services
Other adjacent industries working outside of normal hours

Construction
Multi/single storey/other
Are the walls/roof easily penetrated
Number of entry/exit points in building

Boundary/buildings overview. Comments

Contents

Risk grade

Transport by
Lorry/car/people
Other

Storage of valuables and stock
Method of storage
Storage in vehicles
Other

Ease of transporting goods
Size/weight in relation to value
Volume of goods easily taken by hand
in transportable containers

Contents value
By room
Total area

Control of valuables and stock
How is access limited to staff
Is access limited to authorised persons

Room/area use
Percentage of area use
Shift patterns
Times of area use

Contents overview. Comments

Table 1.3   *Risk level priority chart*

Risk level	Priority	Action
XX	Unacceptable security risk	Immediate attention required to remove risk or hazard
X	Urgent security risk	Must receive attention as soon as possible to reduce or remove the risk or hazard
Y1	Priority	Receive attention to establish if risk or hazard can be reduced and satisfactory procedures put in place
Y2	Low priority	Receive attention at earliest opportunity
Y3	Very low priority	Requires attention when able to

When the **ABC** of risk has been determined it becomes possible to involve all persons who may have an influence in the risk assessment. This can finalize the risk to an **ABC(P)** level where **P** = **people**.

Assessing people can be complex but within this we need to consider:

P (people) $\Longrightarrow$ **Staff history Turnover of staff Vetting information**
Roles of duty and access to the stock and valuables

When all of the assessments have been completed it is possible to make a chart for the risk level to prioritize the action required. Table 1.3 refers.

The surveyor can then overview all of the data and establish the exact grade of risk and determine the costs and the customer/insurance requirements.

It is important not to overlook such things as fire, health and safety breeches of confidentiality and vandalism.

Remember that the security is ultimately graded as a risk from 1 to 4.
This risk will be essentially as Table 1.2.
A full assessment of the risk must always be made in relation to the premises in order to prove the grade as this is a fundamental requirement of the EN.
Log all of the findings from the risk assessment.

As a follow on, the next thing to consider in the risk analysis is the minimum levels of supervision that are needed for the protected premises.

Table 1.4   *Minimum levels of supervision*

Perimeter	Grade 1	Grade 2	Grade 3	Grade 4
Doors	O	OP	OP	OP
Windows		O	OP	OP
Other openings		O	OP	OP
Walls				P
Ceilings/roofs				P
Floors				P
Room	T	T	T	T
High risk objects			S	S

O = opening; OP = opening perimeter; P = penetration; T = trap; S = special considerations

Minimum levels of supervision

Having determined the risk level of the premises, building location and design we can then turn our attention to the minimum supervision levels that are required within EN50131.

The supervision relates to the extent and type of intruder detection devices and protection that is to be installed such as the detection of doors or windows becoming opened plus any attempts at penetration through walls, ceiling or floors etc. These all have to be determined in response to the premises and the contents held within it. Table 1.4 can be used as a guide. This illustrates how perimeter openings are to be protected from Grade 2 upwards and that every grade needs at least one trap technique such as that afforded by movement detectors to give appropriate volumetric protection.

The important thing to remember is that this table is only a guide. It is there to draw the surveyor to areas that could perhaps be critical and to respond appropriately. It is also useful as a means of illustrating just how the levels of supervision change as we move up through the grades.

'Supervision' relates to the detection techniques and the devices employed but for practical reasons only guidance can be given within the EN Standard. This involves surveying for the detection devices to be employed on such openings as doors and windows, trap by movement detection and the sensing of penetration through the building fabric.

As we reach the higher Grade 4 there is a need for penetration detection devices to be installed on walls, ceilings and floors etc. with comprehensive trap and opening detection. However it will be noted that supervision for Grade 1 is limited to only those openings of specific risk and only marginal trap protection is needed.

Therefore for an installation risk assessed as Grade 1 there is no need to employ high degrees of detection, which means the installation can be very much limited in terms of man hours and system interconnections

Table 1.5 *Minimum levels of supervision. Comparison Grades 1 and 2*

Considerations	Grade 1	Grade 2
Doors	**Opening**	Opening perimeter
Windows		Opening
Other opening sections		Opening
Room	**Trap**	Trap

Table 1.6 *Tamper detection. Comparison Grades 1 and 2*

Tamper detection	Grade 1	Grade 2
Control & indicating equipment (CIE)*	✓	✓
Ancillary control equipment (ACE)*	✓	✓
Alarm transmission equipment (ATE)*	✓	✓
Power supply units (PSU)*	✓	✓
Detection devices	✕	✓
Junction boxes	✕	✓

*equipment issue

with a huge reduction in fitting costs, final system complexity and domestic disruption.

Also using on board control equipment, only complying with Grade 1, means that remote keypads with their inherently higher levels of security and generally associated with Grade 2 onwards do not need to be fitted.

So as an example let us consider the recommendations within EN50131 for the minimum levels of supervision of Grade 1. The best way to do this is to do a comparison with Grade 2 because it is Grade 2 that suggests a traditional mainstream British Standard and time honoured intruder alarm. See Tables 1.5 and 1.6.

It follows that the system complexity and inherent cost can be easily controlled for those systems that are risk assessed as being to Grade 1 because the requirements are far less demanding than those of Grade 2. Therefore system setting, part setting and other operational functions become much more understandable to both the younger and older generation of system users. The infirm and nervous user also benefit from a less complex network.

In terms of power supplies Grade 1 and 2 systems are actually the same (12 hour standby capacity) but this as we know is an equipment consideration and Grade 1 systems of themselves tend to have a lesser power consumption.

In addition, event recording is optional with no retention time.

With regard to sounders or signalling, Grade 1 systems only need two warning devices or a traditional self powered warning device. An option can be a signalling system with 24 hour reporting. Basic plug-in speech diallers to the telephone line can easily be added as an extra. The traditional self powered warning device mounted in a prominent external position is a huge deterrent in itself and gives no portrayal of the low risk system grade or the level of protection.

Remember that the system grade is that of the lowest sub grade so various graded components can be mixed in any installation.

Table 1.5 indicates that for Grade 1 the installer need only show an interest in, and subsequently only protect, doors that are *considered a specific risk* or any *particular trap positions*. Other doors and areas do not need protection.

Table 1.6 shows where tamper detection is to be applied as this has a big influence on the supervision and on the wiring and terminating. Both Grade 1 and Grade 2 systems allow tampers to be user reset.

Having been through all of the risk assessment schedules and having established the minimum levels of supervision it is possible to compile a summary security risk assessment form. An example is given as Table 1.7.

1.4   Standards, codes of practice and regulations

In practice there are many standards, codes of practice and regulations that the engineer must become familiar with and use as the base documents when selecting and subsequently installing and maintaining any intruder system. In addition many documents exist which include supplementary information due to changes within this continuously developing industry.

Throughout the course of this book all of the standards, codes of practice and regulations that apply to any given aspect are noted. We have already considered the fundamentals of EN50131 and at a later point, in Section 10.4 the most important documents are listed as reference information. However at this stage it is appropriate that we do an overview of the subject in order to illustrate how certain different specifications have had a control over the intruder alarm industry over the years. It should be understood that we are also about to refer to a number of recently withdrawn standards but the scope of these must be appreciated because their original duty and role will continue to have an effect on the industry for quite a few years.

BS 4737

Although this has now been withdrawn it will still have a certain influence in the industry for some time because it was used as the essential standard for many years. For this reason a huge volume of installations and products were manufactured and installed to meet its guidelines during the time of its formal existence. Although it has been superseded by the EN Standards there was always to be a term of parallel running in the early stages, therefore BS 4737 should still at the least be understood.

Table 1.7   *Summary security risk assessment chart*

SUMMARY.   SECURITY RISK ASSESSMENT

Client _____   Date of risk assessment _____

Person seen _____   Position _____

Address of premises _____

Trade _____   **Risk Grade   1   2   3   4   EN50131**

Type of stock or property to be protected _____

Value _____   Maximum loss possible _____

If client refused to disclose identify tick here ___

Description of premises _____   **Crime rate   V. high   High   Low**

Physical security   No obvious defects   Requires attention
External doors   Locks
_____   _____ ___   Windows
_____   _____ ___   Shutters
_____   _____ ___   Cellar flaps
_____   _____ ___   Other (state)
_____   _____ ___

Occupation: Business   Live on premises   Domestic   Other

Previous history break in   Week   Month   Year   3 years

Entry through   Door   Window   Wall Roof   Other _____

Amount of loss £ ____   Client not prepared to give details _____

Expected response to alarm: Police   Keyholder   Security company

Method of signalling   Audible   Remote (type)

Method of confirmation: Audio   Video   Sequential

Method of unsetting: Unlocking the entry door unsets intruder alarm

Unlocking the entry door disables confirmation

Opening the entry door disables all confirmation

Unsetting by portable Ancillary control equipment

Unsetting carried out with Alarm Receiving Centre

**The intruder alarm system is to be installed in accordance with
EN50131-1 as Grade** ___

See Form – for minimum levels of supervision

BS 4737 was always the essential document for installers of intruder alarms.

Part 1 gave assistance to subscribers, alarm installation companies, insurers and the police in order to achieve a complete and accurate specification of the required protection for audible or remote signalling systems.

Part 2 referred to deliberately operated systems including hold up and bandit alarms, whilst Part 3, which was in various sections covered the different component specifications and continuous wiring used as detection devices.

The code of practice for the planning and installation of intruder and deliberately operated systems was given in one section of Part 4 to cover those systems comprising Parts 1 and 2. Maintenance and records with the action to be taken in the event of false alarms was contained in a further section of Part 4.

The terms and symbols for diagrams were covered by Part 5.

It was the case that BS 4737 covered a mix of equipment, installation requirements and equipment specifications so it did have a great influence throughout its life.

BS 7042

Alongside BS 4737 this was superseded by EN50131. BS 7042 was originally specific to the requirements of intruder alarms for high security systems.

Compliance with it required the installation of special equipment with supporting documents being obtained from the manufacturers.

BS 6707

This is the specification for intruder alarm systems for consumer installation which caters for self-contained units, kits and PA devices intended for the do it yourself market. It gives advice for the planning, installation and use of those systems purchased in kit form. This kit must contain sensors and detectors, warning devices, control equipment with tamper detection and circuitry. Battery only systems must incorporate a voltage test facility or battery state indicator. Self-contained units must have all of these in one enclosure.

This standard was never taken seriously by the professional intruder alarm company and when the ENs were introduced with an official recognition of Grade 1 systems which can also cover components systems supplied as kits the role of BS 6707 became rather insignificant.

BS 6799

This standard was also taken over by the introduction of EN50131 because the latter standard included systems with wire-free links. BS 6799 was originally applied in the industry as a code of practice for wire-free intruder alarm systems. It contained the criteria for the construction, installation and operation of intruder alarm systems in buildings using wire-free links between components such as radio frequency (RF) or ultrasonic connections.

This standard covered systems consisting of one or more transmitter units associated with a detector or detectors and a common receiver unit. It identified six levels of system performance in ascending order of sophistication, from Class 1 to Class 6. Only Class 3, 4 and 5 systems were acceptable for use in audible only intruder alarm systems whilst Class 6 could be used for those installations that required a police response.

BS 6800

This was developed over the years as the specification for home and personal security devices. It detailed the performance criteria for simple security devices that gave an indication of warning when activated. Typical types were video recorder protection devices, door chain alarms, simple access control devices, movement detectors, door glass break sensors, pressure mats, random switching devices, light-sensitive devices and programmable lighting units. All of these products were intended for use separately as individual items and not as parts of interconnected systems. Therefore it had little interest to the professional installer. Its aim as a British Standard was to provide users with a product that had an acceptable level of security consistent with an economical price and to guide manufacturers towards the consumer expectations of such products.

It follows that all installers of intruder alarms are required to install products in accordance with relevant standards, codes of practice and regulations. There are also other documents that may apply and be published by such bodies as the inspectorates or via the police forces. These publications follow certain patterns:

Standards tend to be specifications that relate to equipment and include tests intended to prove that the equipment complies with the standard.

Codes of practice recommend good practice so that the work is performed by competent persons. These documents may be published by an official body, such as the British Standards Institution, or by recognized inspectorates or trade associations to cover a specific area of industry or procedure. Therefore codes of practice relate very often to practices revolving around system design, the installation plus the maintenance and repair. European standards appearing in the form of Codes of Practice are known as Application Guidelines.

Published documents (PD) are in the main formed from supplementary information on aspects of standardization that is not present in a particular standard. They may include material to be used in the interim period because a full range of standards has not been finalized.

Drafts for development (DD) are published on a provisional basis when guidance on a subject is needed. They are eventually issued as standards or withdrawn. DD 243, which has already been mentioned, is important in this respect.

European standards appearing in the form of drafts for development are known as pre-standards and designated 'ENV'.

In order to ensure that consistency is applied within the installation company as to the application of standards it is policy to develop their procedures into a quality management system (QMS). This quality system should be within the framework of the BS EN ISO 9000 series of standards but must depend on the size of the company and the level of the work it undertakes. In many instances it will also be governed by the require-ments of any inspectorate body to which the installer subscribes. BS EN ISO 9001: 2000 defines quality management principles that can enable a firm to be systematic in business activities and improve efficiency. Any firm assessed to the exact requirements of this particular standard as a compliant company can use the 'crown and tick' logo and this of itself can have a huge impact on the generation of new business for those wishing to trade in the corporation and local authority sector.

However, it is to be understood that many companies although sub-scribing to the basis of the BS EN 9000 series for their quality practices do not need to develop and operate a specific QMS to be enrolled as a regis-tered installer under every inspectorate scheme. The extent therefore to which the BS 9000 series is applied varies enormously within the different installation companies.

Although we have come to understand the application of the standards that govern the UK and have some influence in Europe, there are many differences in the wider world. Certainly BS 4737 had some alignment with International Publication IEC 328 produced by the International Electro-technical Commission (IEC) and had a measure of acceptance in countries that adopt IEC Standards. However, North American standards differ. This should be appreciated because of the acceptance of many North American policies in the developing world.

In the USA, Underwriters Laboratories (UL) operates approval schemes that extend into the security manufacturing area. UL has strong links with the insurance industry and operates as an independent non profit making corporation to test products for public safety and its service is used by insurance companies as a means of assessing risks to be covered when providing policies. The influence of UL approved products used in secur-ity systems also extends into those countries that respect North American policies and practices. It follows that there will always be a range of stand-ards to which the intruder alarm engineer must subscribe in the inter-national scheme of things. However at present we are concentrating on the impact of British Standards.

Codes of practice will always follow British Standards with influences from such policies as implemented by the ACPO. These will then be adopted as base documents by inspectorate bodies or authorities such as the Loss Prevention Council (LPC). As an extension on this theme, BS 7671, *Requirements for Electrical Installations*, is issued by the Institution of Electrical Engineers as the IEE Wiring Regulations.

It is agreed that under any standard or code of practice when dealing with installed systems maintenance should be inherent. Throughout this book we will consider the technical aspects of maintenance, servicing and customer care, but now we only need summarize the key requirements of and for a maintenance schedule:

- An intruder alarm is designed to involve not only the customer but many other parties. In this context we include neighbours, the police, insurance agents, keyholders and the installation company. All of these influence each other. It is vital that all reasonable measures have been employed to ensure the correct function of the system since the alarm has an impact on all of the mentioned parties.
- The police insist that false alarms be kept to a minimum.
- The insurers insist that all reasonable steps have been taken to safeguard the property.
- The keyholders have a right to expect no disruption.

Accepting that the alarm system is correctly maintained is never the total answer but it does remove negligence in terms of the requirements of the standards. It also takes away what could otherwise be very costly call out charges which may well be greater than if there is no formal maintenance agreement.

The customer should be informed that a non maintained system, including bells only and Grade 1 is in breach of the ENs and advise of the consequences with regard to the police, insurers, keyholders, the installer and the regulatory bodies.

EMC Directive and CE marking

The EMC Directive is relevant to all types of electronic equipment including electronic security products, and extends also to video recorders, television receivers, computers, hi-fis, vacuum cleaners, washing machines, microwave ovens and suchlike.

The EMC Directive was introduced to ensure that electronic products sold within the European Union or any other countries do not cause excessive electromagnetic interference nor are unduly affected by it.

This interference refers to any electromagnetic disturbance or phenomenon which may degrade the performance of a device, control equipment or system. This disturbance can be in the form of electromagnetic noise, a signal or a change in the propagation medium itself which can contribute to equipment malfunction.

The most important item relating to interference within most electronic products with which we have an interest is the microprocessor.

The CE mark C€ is applied by the manufacturer to signify that a product conforms to a particular European regulation. In the case of intruder alarms the application of the mark denotes conformance with the European EMC Directive. It can be found on the apparatus itself, the packaging, installation instructions or on a certificate of guarantee accompanying the goods.

Effective date

Products manufactured on or after 1 January 1996 are required by law to conform to the directive and to carry the CE mark.

The effect on the installation of electronic security equipment

Any installation after the effective date requires CE marked products in cases where the equipment has been manufactured on or after the effective date. This also applies to product replacement within existing installations.

Affect on stock

Installers holding stock manufactured prior to the effective date will find that items are not necessarily CE marked. However, they may still be legally used as a part of a new installation if identified as being manufactured prior to this date.

Assurance of EMC conformance

For any product featuring the CE mark, the manufacturer is responsible for issuing a declaration of conformity to document compliance with the directive. This declaration must be made available to any customer on demand. In addition, the producer of the goods must be able to give evidence to support the declaration of conformity – usually in the form of a test report.

In the case of products exhibiting transmitter/receiver characteristics such as microwave or dual-technology detectors this testing must be performed by an accredited test house. The report must then itself be verified by the Radiocommunications Agency before EMC Directive compliance can be claimed. For other products, the manufacturer can self-certify by doing in-house tests at the point of production. In these instances the product is marked CE0192.

Liabilities

The manufacturer is responsible for ensuring compliance. Failure to do so can result in a fine, imprisonment and the recall of non-compliant goods.

The installer is obliged to fit the goods in accordance with the manufac-turer's instructions. Non-compliance can lead to the same penalty as that for a manufacturer.

1.5   ACPO Security Systems Policy and response organizations

Originally the local police force would issue a 'Police Recognized List' for a given area which was a subscriber list for the public providing information about companies that could be relied upon to carry out a neat, cost-effective and correct installation to the then BS 4737. This concept was withdrawn following the introduction of a new Unified Intruder Alarm Policy and in keeping with developments in the market for security systems. In 1995 the Association of Chief Police Officers (ACPO) revised its policy on police response to alarms and this was fully implemented on 1 April 1996. It became known as the 1995 ACPO Policy and was effectively the NACOSS code of practice NACP 14 or its equivalent industry code of practice SCOP 14. It stated:

> The aim of this policy is to enable the police to provide an effective response to genuine intruder alarm activations thereby leading to the arrest of offenders and a reduction of losses by improving the effect-iveness of alarm systems and reducing the number of false calls to the police.

This policy was quite rigid in most of its requirements, but it only applied to systems with remote signalling with a police response. Audible-only systems did not fall within its scope. It reinforced the need for engineer reset on remote signalling systems with a police response although reset-ting could be performed by the central station if sufficient information was available. Certainly the policy placed the responsibility for choosing an alarm company directly on the customer as the police would only respond to alarm systems which had been given a unique reference num-ber (URN). For a company to be able to issue URNs it must be recognized by an inspectorate body. Following on from this the ACPO Security Systems Policy 2000 was introduced. It applies to all sites, whether domestic dwellings or business, which employ electronic security systems and require the police to attend if an alarm condition is generated. The new policy once again accepted a number of organizations for the man-agement of false alarm procedures, thereby allowing installers of those organizations to install police-calling alarm systems by issuing URNs.

Under the new ACPO policy, inspection bodies approving security installers who install police-calling alarm systems also need to be accredited by the United Kingdom Accreditation Service (UKAS), who operate on behalf of the Government under a Memorandum of Understanding from the Department of Trade and Industry. The new policy also made changes to the police response depending on the number of false alarms in any rolling 12 month period, in effect tightening the rules on the managing of unwanted alarm activations.

All alarm systems issued with a URN are automatically entitled to Level 1 immediate response but this is reduced to Level 2 if the police are called to two false activations in a rolling 12 month period. If problems persist and the system is further reduced to Level 3 the police response is effectively withdrawn. In such circumstances, Level 3 will continue to apply until the system is free of false activations for three months, but if this is not achieved within a period of six months the URN will be deleted.

Systems which include a PA (personal attack) facility should be capable of sending a signal which differentiates between PA and intruder alarm generation. If not, a reduction in police response will include all elements of the system. It is to be understood that PA units should only be used when an individual is at risk from attack, and therefore when they signal separately and distinctly they can be viewed as systems in their own right.

The requirements of the new policy can be compared to that of the original over a rolling 12 months:

ACPO Security Systems Policy 2000	Old policy	
1–2 false calls	1–4 false calls	**Level 1** *Immediate response*
After 2 false calls	After 4 false calls	**Level 2** *Police attendance may be delayed due to other priorities*
After 5 false calls	After 7 false calls	**Level 3** *Police response withdrawn* *Keyholder notified only*

←———————— Rolling 12 month period ————————→

The police response is defined as the police attending the protected premises when they are informed of an alarm activation. The installed system must be classified under the policy as Type A or Type B.

● *Type A:* remote signalling systems. These signal alarm activity to either a BS 5797 recognized alarm receiving centre or to a police control room with the specific approval of the Chief Officer of Police. These systems also need to meet the necessary British Standards or industry Codes of Practice and be installed, maintained and monitored by companies certified by a UKAS accredited certification body. If not the system is unable to be classed as Type A. All Type A systems are issued with a URN, which is necessary in order to qualify for police response.

Type A systems need to have two designated keyholders, who need to be trained in the use of the alarm system, be telephone subscribers and have transport if necessary, enabling them to be at the site within 20 minutes of being notified.

- **Type B:** security systems. These are security systems for which a police response may be requested but do not conform with the guidelines for Type A systems. Type B systems will not be issued with URNs and therefore do not receive police attendance automatically.

Systems need to be Type A as designated by the policy, which means that they must signal to a recognized manned station. A new perspective of the policy is that this signalling must incorporate confirmed activations. The techniques associated with this are covered in Section 6.9, Alarm confirmation technology. In practice these confirmed alarm techniques use a second activation to confirm that the initial alarm activation is genuine. DD 243: 2004 defines the possible procedures for transmitting these confirmed alarms through to the ARC.

It is achieved by one of three established means; namely visual confirmation, audible confirmation or sequential confirmation. For visual and audible confirmation additional equipment is required to provide the alarm receiving centre with the confirmation signal. Sequential confirmation is the simplest and most adopted technique and is used to indicate that a previously signalled alarm condition is likely to be genuine. It is achieved through a 'second alarm' signal sent when a separate zone to the one that caused the initial alarm is activated. As an example, an intruder who opened a door would initially trigger a 'code 3' alarm and when he subsequently triggers a movement detector the 'second alarm' is generated and the ARC is able to action a confirmed alarm to the police. The transmission procedures are not applicable between two detectors covering the same area as the environmental problems triggering one could have the same effect on the other. Therefore two detectors of the same technology cannot cover the same area, or if they do they must use different techniques (e.g. PIR and microwave).

Alarm confirmation is a system parameter rather than a panel feature so the site always needs enough detectors to ensure appropriate coverage in order to generate the second alarm condition. DD 243 defines the possible procedures for transmitting a confirmed alarm through to the ARC:

- The initial alarm condition can be sent to the ARC followed by the second alarm signal.
- A single signal can be sent to the ARC indicating that a confirmed alarm has been detected.

These methods are specified in the code of practice DD 243: 2004 which gives guidance and advice on the design, installation and configuration of intruder alarm systems which signal confirmed activations to an ARC. It does not cover PA elements but does not imply that such systems cannot incorporate PAs. All systems which apply for an initial URN after 1 October 2001 must be capable of creating confirmed activations as specified in DD 243 or if it has been removed from a Level 1 police response (immediate) and wishes to be reinstated to that level. As an addition, if a signalling path has the capability to signal a loss of communication (dual signalling systems), a communications loss followed by an alarm, an alarm followed by a communications loss, or two communication losses on separate signalling paths, this may also count as a confirmed activation.

The intention of the policy is to reduce false call-outs for the police and this has an obvious knock-on effect for installers. Although they will not have to replace a complete system they will have to ensure that all alarms within the proviso can be confirmed. The real perspective of ACPO 2000 remains that all end users' properties are adequately protected and that fewer false alarms are generated. In practice, DD 243 in its role within the policy only applies to intruder alarm systems and does not limit the role of the installer in the creation of the system. Indeed it offers alternatives to cater for diverse applications. The areas covered are:

- *Primary design and configuration considerations.* This relates to design objectives, the acceptance of the different alarm confirmation technologies and the transmission fault signals in confirmed activations. It also covers the need for providing information to the ARC as regards the level or changes to the police response.
- *Design, installation and configuration of intruder alarm systems which incorporate confirmation technology.* This deals with design considerations and the provision of effective confirmation technology and the notification of alarms. It also involves the location of control and indicating equipment and how systems may be configured.
- *Handling of information by ARCs.* This is related to the handling of information by the ARCs so that the operators follow the code of practice in order that the signals are classified as confirmed.
- *Minimization of false alarms.* This is of immense significance so that false alarms may be filtered out and includes the means of setting and unsetting the system.

False alarm prevention

False alarms are caused by all manner of circumstances. Much emphasis is therefore to be placed on careful system design and installation practices, as well as the selection of appropriate equipment to meet the needs of a given site. ACPO are determined to reduce the incidence of false alarms that are caused when a user enters the protected premises so we expect to see unsetting in the long term further biased towards electrically contacted locks which, on turning a key, unlock the door and unset the system (Block-schloss). Tags of course also help to reduce false alarms on unsetting as only 'one single consistent action' is required. At present the ACPO policy accepts a number of unsetting practices but a popular technique is by opening the entry door and then turning off the system by a key fob or card swipe – not by a keypad. The acceptable practices of unsetting to DD 243 are covered at a later point in this section.

Overall we can say that the largest percentage of false alarm activations are actually caused by user error but there are some basic rules that apply to all intruder systems in general. Installers and clients should appreciate the following points:

- At the outset inform the system users of the consequences of false alarms. Clearly indicate their responsibilities.
- Realize that the installation company is responsible for false alarm performance.
- Ensure that the alarm receiving centre accepts a measure of responsibility for the collection of alarm signalling data.
- Ensure that any animals or pets are unable to enter areas that are fitted with detectors unless purpose-designed sensors are used.
- Confirm that detection devices are not obstructed.
- If movement detectors are installed do not introduce new sources of heat, movement or sound in the areas they protect.
- Confirm that all doors and windows that feature contacts are closed in advance of setting the alarm.
- Never deviate from the designated entry/exit routes when the system is set.
- Have all keys, codes and fobs or cards available when entering or leaving the premises.
- Ensure that all system users have been correctly trained and minimize the number of people who operate the system.
- Ensure that all users understand the official procedures for any abort process required if incorrect codes are entered into the system.
- Never disclose codes or passwords to unauthorized persons.
- If the alarm system control panel features a fault message for a specific zone when setting, check that the detector is not being triggered by a spurious event and that doors where protected are closed.
- If faults cannot be cleared in any zone contact the installation company.
- Ensure that a maintenance programme is followed.
- Clients are not to tamper with or attempt to move any part of the alarm system.
- The installation company should be advised of any alterations that may affect the alarm system or any damage caused to the system, its detectors or wiring.

In order to reduce false alarms and in recognition that the majority of these are caused within 90 seconds from an alarm set, all systems must be able to transmit either an open/close or an abort signal. This means that the system needs to be capable of indicating to the ARC that the system is set or unset or it can generate a secondary signal to show misoperation. Internal warnings are needed close to the setting point so that on set a key-holder can, if necessary, return to the panel and enter a PIN to notify the central station via the communication device that no action is necessary if a setting fault has occurred. Therefore in the design the intruder system must incorporate a technique to signal to the ARC whether it is set or unset and be capable of sending a secondary signal in the event of user error occurring. This is in addition to designing a system to giving confirmed activations from the detection devices that in the case of sequential confirmation consists of two separate alarms being processed within a specific time window.

In order that the confirmed activation cannot be caused by user error, the system must of itself include specific setting and unsetting procedures. These are typically:

Setting

- Setting via a key-operated shunt lock at the final exit point.
- Setting by a push button outside of the protected premises.
- Setting via a door contact or similar protective switch at the final exit point.
- Setting by a portable device ACE (ancillary control equipment).
- Remote setting by the ARC.

Unsetting

Unlocking the initial entry door to unset the whole or unset part of the system facilitating the following:

- in part set, entry to areas to be prevented by locks
- unlocking the doors to set areas unsets that part of the system
- status of locks to be monitored at the control panel
- system will be prohibited to set until all relevant doors are locked.

Unlocking the initial entry door to disable all means of confirmation
If the entry timer expires it will be signalled to the ARC as an unconfirmed alarm. Any deviation from the entry route will be signalled as an unconfirmed alarm.

Completion of unsetting using portable ACE
This can be done with a key fob but the entry door must still be fitted with an entry timer. Alternatively, opening the entry door starts the entry timer and the system is then unset by a key fob or other token. If an alarm occurs

during entry or as a result of the expiry of the entry time the alarm is unconfirmed. Using this method of unsetting a confirmed alarm can occur if two or more independent detectors located off the entry route are activated after expiry of the entry time.

Remote unsetting by the ARC
This can be done using downloading features of the system using specific software.

It is to be noted that unique terminals on the control equipment are needed for many setting/unsetting practices and that the manufacturers' data will provide information on the programming required. In addition, when protective switches are used for setting, such as a door contact, the control equipment is to be located near to the final exit point so that a secondary misoperation signal may be sent quickly in the event of user error. A set signal is to be sent to the alarm receiving centre and an internal warning should follow at the protected premises so that false activations may be avoided.

It remains to say that contrary to common opinion the police will only attend sites in response to an alarm system if steps have been taken, as best as is practicable, that genuine alarm activations are passed to them. The policy also accepts a number of organizations for the management of false alarm procedures, thereby allowing installers of those organizations to install police-calling alarm systems by issuing unique reference numbers. There are a number of schemes and inspectorate bodies prominent in this role.

Inspectorate bodies

Under the ACPO policy the inspectorate or inspection bodies who approve the security installers that fit the actual police-calling alarm systems need to be accredited by the United Kingdom Accreditation Service (UKAS). By achieving this EN45011 accreditation the inspectorate meet the revised ACPO intruder alarm policy as well as the requirements of the Association of British Insurers. In this role it is UKAS who operate on behalf of the government under a Memorandum of Understanding from the Department of Trade and Industry.

In practice UKAS accredits the inspectorate bodies who themselves provide the specific inspection and approval schemes for the installers across the various sectors of the security industry. Such inspection bodies may be formed from a number of organizations with a divisional structure that deals with the different electronic security systems or manned services.

As the intruder industry has developed over the years the inspectorate bodies that have provided the services have changed somewhat. At the present time the intruder alarm engineer is interested in the National Security Inspectorate (NSI) and Security Systems and Alarms Inspection Board (SSAIB) schemes.

Police response

In order to qualify for police response for monitored intruder alarm installations, under the ACPO and ACPOS (Scotland) policy, there is a need for installers to be approved by NSI or SSAIB.

For monitored intruder alarm systems the police will only offer a response to a verified alarm system which has been granted a police unique reference number (URN) and has been routed via an alarm receiving centre as endorsed by the ACPO Unified Policies.

Security installers covered by the NSI and SSAIB schemes comply with these requirements so their systems do qualify for police URNs.

In all other respects the police will continue to respond to 999 calls as appropriate.

In practice these two bodies are structured in the industry to embrace more than intruder alarms since they extend their services using different divisions to cater for all of the mainstream security disciplines to include fire detection, guarding/cash services and ARCs etc. However we will concentrate our efforts on how they actually approach intruder alarms. Remember that both the NSI and SSAIB are accredited by the United Kingdom Accreditation Service (UKAS).

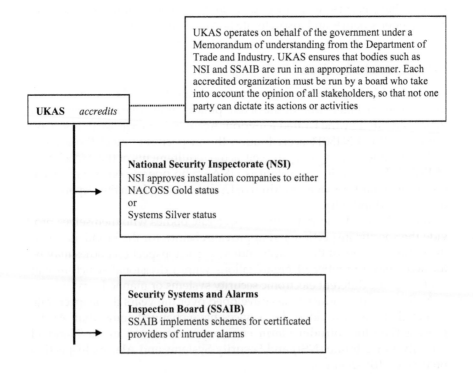

The thing to remember is that by operating UKAS accredited schemes NSI and SSAIB provide the mechanisms by which providers of such services as security, fire and guarding can demonstrate their competence and ability to operate well managed and professional businesses.

The Quality Management and Product Certification schemes accredited by UKAS are represented by specific logos.

National Security Inspectorate (NSI)

This is the national body that inspects firms providing security services across the United Kingdom and beyond. It covers electronic security and manned services and oversees other areas related to security and safety. It was launched to strengthen other already well established schemes at the time.

NSI is an independent, not for profit approvals body providing inspection services for the security and fire industries. For many years it has been protecting the customer through insisting on the highest standards and operating the toughest inspection regimes. Companies are inspected regularly by highly qualified, full time inspectors to prove their ongoing competence. NSI's customer care support process reinforces the reliability and integrity of the companies it inspects. All NSI schemes meet or exceed the appropriate requirements of the police, fire services, government, industry and insurers.

The mission of NSI is to raise standards in the security and fire industries in the interests of the customer and to achieve recognition as the benchmark of public confidence in the security and fire industries.

The police, fire services, insurers and other response organizations trust NSI on the basis of:

- **Quality delivery.** Highest possible performance standards in the UK with onerous inspection regimes.
- **Security vetting.** Security staff are strictly vetted.
- **Training.** High level of staff training and supervision.
- **Maintenance and call-out.** Companies have to deliver quality service/ maintenance (e.g. 24/7 4-hour response).
- **Police response.** NSI approved security systems companies are authorized to install police-calling systems and NSI companies operate highly efficient police-calling processes and networks in the UK.
- **False alarms.** NSI approved companies are required to design systems to minimize false alarms.
- **Deterrence.** Anecdotal surveys show a high criminal awareness and avoidance of premises protected by NSI approved companies.
- **Customer care.** NSI offers an independent complaints arbitration service.

NSI operate three tiers of approval – *Gold, Silver and Bronze.*

These recognize that companies are at different stages of development, providing smaller and newer companies with an opportunity to receive an appropriate level of approval on a ladder of performance.

The Gold schemes are the most esteemed and provide a target for ambitious companies to achieve. The Bronze scheme is an entry level scheme only for guarding companies, which must then achieve Silver, within a given time scale. Therefore intruder alarms are only covered by Gold and Silver levels.

All NSI Gold and Silver approved companies meet the requirements of the police through the ACPO policy and will respond to confirmed alarm activations for monitored systems.

Insurers will often insist on an NSI NACOSS (National Approvals Council for Security Systems) Gold or Systems Silver approved company being used subject to the level of risk. For high risk, insurers typically require NACOSS Gold.

NACOSS
GOLD

SYSTEMS
SILVER

There are certain differences between all approved NSI companies and those with Gold status:

All NSI approved companies:	NSI Gold companies also:
• Meet the most rigorous entry standards in the UK industry. • Are subject to comprehensive 'Repute of Management' checks. • Meet all relevant British and European Standards for technical performance. • Have received a 10 year security vetting for staff (security companies). • Must train staff to industry best-practice standards. • Are fully insured to protect customer interests and staff. • Are inspected every six months by professional auditors including a sample.	• Meet the industry-specific ISO 9000 Quality Management System. • Require additional tougher entry standards and due diligence on application and ongoing inspection. • Operate formal, inspected internal procedures and audit processes to give greater customer assurance. • Have a long term track record of performance and evidence of reliability and stability. • Provide excellence in business efficiency and customer care. • Offer a high level of insurance specification and preferment.

Gold companies will be subject to two annual inspection visits. These are typically:

First annual inspection visit
Office focused, looking at management procedures, document control, customer care, training, false alarm management and security screening of new employees.

Second annual inspection visit
Site based and selected randomly by the inspectors to check the quality of the work carried out, which may include going back to follow-up any outstanding issues from previous visits.

Inspections always commence with an opening meeting with relevant company employees to discuss the objectives of the day, site visits to be conducted and any matters outstanding from previous visits.

A broad spectrum of site visits is always carried out so these may range from domestic installations through small warehouses and offices to large monitored installations.

Following each site inspection the inspector will review the checklist to identify how many, if any, deviations have been recorded and issue an 'audit observation' to the company.

This will be graded A to E where:

A means there are no deviations
$B = 1–2$
$C = 3–4$
$D = 5–6$
$E = 8+$

Where B or C is recorded, the company's documentation will be checked at the next visit to confirm the corrective action was completed within the required time.

D and E grades will require re-inspection within 3 months to confirm that the work has been completed to standard.

In all cases the NSI inspectors spend time listening to company issues as this provides essential feedback.

Security Systems and Alarms Inspection Board (SSAIB)

Founded in 1994, SSAIB is a not-for-profit organization, set up as a company limited by guarantee, with a Governing Board representing a wide range of industry stakeholders. In addition to representation from the police (ACPO and ACP Scotland), fire services and insurers, the Board comprises representatives of various government departments, trade associations, professional bodies, practitioners and relevant consumer organizations.

SSAIB has wide UKAS accreditation, which means that their approach is subject to independent scrutiny and validation against international standards.

As a certification body SSAIB's remit is to ensure that the interests of the consumer (e.g. purchasers, specifiers and end user groups) are protected. This is achieved through a strict enforcement of quality and performance standards.

Electronic security systems vary enormously in capability and complexity and by using an SSAIB certificated provider ensures that the design, installation and maintenance of the system will comply with all relevant standards.

There are a number of product schemes available within SSAIB. Intruder alarms are contained within the Security Systems scheme and offer certification. These are schemes which are independently assessed, accredited and monitored by UKAS. SSAIB also offers ISO 9001: 2000 Schemes designed to complement its product certification schemes. Management systems certification is available for intruder alarms. Effective quality management systems provide a structured and systematic approach to help achieve the company's aims and objectives. It helps to streamline business processes and improve self-governance.

Scheme	Activity	Certification	ISO 9001
Security systems	Intruder alarms	✓	✓

Certification. ✓
Those schemes which are independently assessed, accredited and monitored by UKAS
ISO 9001. ✓
Management systems certification focuses upon the business framework of the company. It provides a structured and systematic approach to help achieve the company's aims and objectives.

Using an SSAIB certificated provider ensures that the design, installation and maintenance of the system will comply with all relevant standards.

All providers must demonstrate that they:

- Install and maintain equipment to relevant British or European standards
- Have appropriate management systems and procedures in place
- Have relevant personnel security screened
- Meet specified financial criteria and be financially stable
- Have appropriate insurance cover
- Have sufficient resources to meet their commitments
- Have suitable premises, equipment and vehicles
- Have trained staff and implement training procedures
- Have sufficient systems installed and/or monitored to allow a trading history to be demonstrated
- Provide 24 hour service and repair cover
- Issue a Certificate of Compliance for the services provided
- Are in a position to carry identity cards

Providers of all of these services are subject to regular inspections to ensure that installations, management and maintenance standards are maintained.

For some installations periodic maintenance may be a condition of cover, for example monitored systems and systems installed for insurance purposes.

Using a third party certificated provider may be a requirement of insurance cover for a commercial application and it may also yield premium discounts for domestic applications. In both cases, SSAIB certificated providers will meet or exceed insurers needs.

SSAIB maintain that for the customers standing that if they are not satisfied with the service received from the certified company in terms of the system failing to comply with the agreed specification SSAIB will investigate the complaint.

If substantiated, SSAIB will require the installer to rectify the cause of the complaint.

The UKAS accredited inspection bodies work to strict criteria laid down for the monitoring of installation companies and their compliance with the relevant European and British Standards and codes of practice for the intruder alarm industry. Therefore, checks are made on installation companies against, not only, sample installations but on records, accounting systems, false alarm management procedures, 24 hour emergency cover, insurance policies, public liability and efficacy insurance. There is also information to be logged pertaining to the installation – detailed specifications, after-sales service agreement, and routine and remedial service sheets with the purpose of each visit.

Therefore, the following questions will always be asked of an alarm installer.

- Is the company in a position to offer 24 hour emergency call-outs together with a maintenance contract?
- Do they have an identification card with a photograph?
- Do the installations meet the relevant European or British Standards?
- Is the installer recognized by a UKAS accredited certification body and therefore able to issue URNs? If not, is the company concerned about being verified by an inspectorate body?
- What is the overall cost of the installation, what exactly does it cover and will it include a routine annual visit?
- Does the installer give written specifications and a copy of the terms and conditions?
- Is the equipment outright purchase or are parts of it lease?
- Is there a manned 24 hour contact line or is there a pager/mobile network service?
- Is it possible to obtain references from past customers?

All of these aspects are to be budgeted for, so cost comparisons against the inspectorates should be established. The cost of information packs, inspection fees, annual fees, certificate costs and additional certificates must be sought. This must then be offset against installation prices as the company is responsible for making a profit. The engineer can then ask what other considerations there are and the time needed for procedural matters and producing documentation.

With some understanding of the role that the inspectorate bodies play we can next consider the trade association within the security industry.

British Security Industry Association (BSIA)

The British Security Industry Association is the professional trade association for the security industry in the UK. Its aim is to help its member companies succeed in an ever-changing and highly competitive business environment.

The membership sections include the range of security systems and services with the Electrical Contractors Association (ECA) forming the alarms division. Its member companies are responsible for more than 70 per cent of UK business, including alarms, CCTV, access control, manned security, shutters, grilles, safes, secure transport and alarm manufacture, distribution and installation.

The BSIA is not an inspectorate or certification body but its key role is to promote its members and indeed the needs of the industry as a whole. This is achieved by reacting to companies' concerns and providing a strong voice on these issues to the Government and other key groups or organizations to achieve any necessary change.

The BSIA is the recognized source of information for the media in the security industry and ensures that its members have current information on areas such as legislation, standards and changes in police policy. It follows that whatever a customer's security requirements, choosing a BSIA member company is a guarantee of a secure product or service. Indeed the BSIA was the first trade association to achieve the BS EN ISO 9000 quality management standard. The adoption of the standard as a condition of BSIA membership allows customers to differentiate between professional companies with a quality culture from those unwilling to invest in developing quality procedures. In addition, the BSIA plays a vital role in managing the inspection bodies, and recognition by the BSIA of the importance of training to the long-term benefit of the industry was instrumental in the creation of the Security Industry Training Organisation (SITO). It can be said that specifiers who select BSIA companies can expect a high level of competence as it is the most effective security grouping in the UK.

A number of changes occurred within SITO as an ongoing review of their range of products and services that they needed to offer to the industry and this led to the formation of Skills for Security.

The key objectives of Skills for Security can be overviewed:

- To understand and identify current and future skills needs, gaps and shortages in the security business sector.
- To work with employers, learning providers and other stakeholders to influence and increase education and training supply and to improve workforce skills.
- To promote recruitment and career paths in the sector to attract and retain the best staff.

- To engage and influence government, devolved administrations and regulatory bodies for the benefit of skills development in the sector.
- To establish and maintain a viable commercial organization, capable of delivering products and services to meet the needs of the industry.

The BSIA focuses on ensuring that legislative developments are appropriate and cannot create unreasonable burdens for its members. Therefore it has a close working relationship with Government officials so that industry views are properly considered. In addition it is in close liaison with ACPO over its security systems policy.

Although the BSIA does not inspect companies to confirm that they meet industry standards it does play an important role in standards development. The Association is equally involved in European standards work and provides its members with extensive guidance on standards developments.

The BSIA also produces technical literature, guidance notes, codes of practice and training materials to reassure customers that member companies are committed to good practice. Of particular interest are the EMC *Guidelines for Installers of Security Systems*, which consider problems of electromagnetic compatibility and how to overcome them. Of equal importance is *Guidelines for the Use of Downloading in the Security Industry*, which helps to establish recommendations for downloading and to assist providers of this service with information to meet specifier or user requirements. In the process of promoting developments and issues, the BSIA has considerable scope for highlighting the achievements of its members, and the benefit to customers of selecting contractors who have chosen to conform to its membership criteria.

Insurance company requirements

It becomes apparent that the installer of intruder alarms must also satisfy the clients' insurance requirements as a condition of cover. This follows on from a risk assessment survey of the premises. It is based on underwriting guidelines and on the premise that alarm protection is necessary in relation to the nature of the business, its location and any history of associated risk. However, the insurance company does rely on the end user to choose the correct installer for the job and for the installer to act professionally in regards to the system design.

In certain cases the risk assessment may warrant measures to be taken that are in excess of those stipulated in traditional standards and policies. In general the technical requirements of the Association of British Insurers (ABI) in relation to alarm confirmation systems in commercial premises quote that:

- Control and signalling equipment must be located in a position where it is concealed from general view and is least vulnerable to attack.

- The alarm system must have an event memory with the capacity to store 200 events (100 if this is sufficient for normal use over 40 days).
- The external warning device must be installed at a height of at least 3 metres, otherwise a second sounder will be needed on another elevation of the building.
- Any delay on local sounders must be automatically removed in the event of any loss of remote signalling capabilities.
- Internal warning devices and set/unset sounders must be sited away from the main control and signalling equipment so as not to identify positioning when activated.
- Detectors incorporated into the final exit route should create an alarm condition if their activation is not preceded by the initiation of the correct unsetting procedure.
- Preference should be given over to the use of Loss Prevention Certification Board approved products if suited to the particular installation.

Insurers want to ensure that areas containing valuables cannot be entered through internal doors or via windows without creating confirmed detection conditions, therefore the building shell must be adequately protected by the first alarm detector. Volumetric space detectors would tend therefore to be used for the second alarm activity.

It is normal also that the insurers do not always expect a keyholder response to an unconfirmed detection provided that the unconfirmed detection does not indicate a perimeter breach and the system can rearm in its entirety. These unconfirmed activations involving devices intended to detect a breach of the building perimeter must nevertheless be presented to the ARC before the expiring of any filtering period.

It will be understood that the intruder alarm industry is developing and changing to embrace new technologies and policies.

The Security Industry Authority (SIA) which was set up by the Private Security Industry Act 2001 to regulate the industry, but originally perceived to pertain mainly to the guarding sector, may now have an influence in licensing of the systems sector. Therefore such proposals and influences will ensure that the quality levels within the intruder alarms sector will continue to rise. For this reason we are seeing ongoing updates to such documents as PD6662 and DD 243. In this book we refer to the later 2004 editions although many installations will be specified against earlier editions.

There are also ongoing recommendations from the BSIA and other interested parts of the industry to change thresholds for false activations. Certainly fully monitored, less complicated user friendly systems will further reduce false alarms. In real terms most false alarms are caused by end user confusion during setting and unsetting and key fobs can help in making the intruder system more easily managed.

It also follows that with the reduction in police forces, alarm administration departments and possible URN/Premium rate phone call revenue that initiatives will always exist. An approach can be to reduce false alarm levels to three for non DD 243 compliant systems only, which would give an incentive to customers to upgrade to DD 243: 2004, retaining police response for five activations.

Therefore the ACPO Policy is constantly under review and a living document is available from the Security Systems working group within the General Policing Business Area in relation to the police response to security systems. It was last reviewed in October 2005 to include interim amendments of September 2005. The BSIA in discussion with ACPO on the issue of current systems agreed:

- All new quoted systems installed from October 2005 need to be installed to PD6662: 2004
- If a BS 4737 unconfirmed system with a URN drops to Level 3 (loses police response) then to regain this response it has to be upgraded to DD 243: 2004 (not PD6662)
- If a confirmed system to BS 4737, DD 243: 2002 with a URN drops to Level 3 and loses police response it must then wait three months or be upgraded to DD 243: 2004 (not PD6662) to regain police response
- If a confirmed system to BS 4737, DD 243: 2004 with a URN drops to Level 3 and loses police response it must then wait three months or have an additional form of confirmation installed to regain police response
- If a PA with a URN drops to Level 3 and loses police response it will have to wait for three months and if still required by the client will need some form of intervention/verification added to the system
- If a BS 4737 system with a URN loses this URN then it will have to be upgraded to PD6662

Therefore we may well see a ratified ACPO Policy appearing as the 2005 edition. In particular the threshold for PAs is always under review and this may well be reduced to two to reflect an even newer policy.

1.6 Local authority requirements

The ACPO Policy was related very much to remote signalling and did not affect the installer involved with local signalling. However, these are governed by the local authority, with the principal interest in the control of noise. In the first instance we should recognize what the term 'noise' actually means, and define it as unwanted sound.

Clean Neighbourhoods and Environment Act

In the first instance we need to appreciate the significance of the Clean Neighbourhoods and Environment Act as this came into effect in England

and Wales from April 2006. It had a major influence on intruder alarms as its aim was to reduce the problems of noise pollution by minimizing false alarms.

The Act allows a local authority to designate all or any part of its area as an 'alarm notification area'. This means that the occupier or owner of any premises, residential or non residential, in the area is obliged to notify the local authority of the contact details of a keyholder for the subject premises.

The authority can then turn to that keyholder for assistance in silencing an alarm. It is an offence to fail to nominate or notify the local authority of the details of a keyholder.

The BSIA has pointed out in its 'Industry Strategy for Intruder Alarm Systems' that this is to be a policy within the Act. It can be said that the easiest means of cutting noise pollution from audible intruder alarms is to abolish the use of external sounders as signalling devices. It would effectively mean that all alarms would become Type A remote signalling but without any external sounders. The owner would know that their property was monitored and that in the event of a false alarm they are not going to be penalized by the local council. As environmental issues become more prevalent and as technology improves it could reduce the burden on owners yet still give similar levels of security. A deterrent effect could be maintained by using appropriate signs and warning notices. It is felt that many households would be happy to upgrade their systems to Type A. In practice this would mean more work for installation companies.

The further part of the legislation states that the local authority must be informed in writing of the name, address and telephone number of the nominated keyholder.

Nominated keyholder

The responsible person is the occupier of the premises, a person entitled to occupy the premises or the owner of the premises. The keyholder must be nominated either within 28 days of the 'alarm notification area' designation coming into effect if the intruder alarm was already installed or within 28 days of an installation being completed if the designation is already in effect.

A keyholder can only be nominated if that individual:

- Holds keys sufficient to enable them to gain access to the part of the building in which the alarm controls are sited
- Normally resides or is situated within the vicinity of the premises
- Has sufficient information to enable them to silence the alarm; and
- Agrees to be the nominated keyholder

Keyholders can be key holding companies if they are able to be contacted at any hour of the day. In addition, the local authorities are allowed to set fixed penalties for their areas.

These fixed penalties can be applied when a keyholder is not nominated or details are not notified to the local authority.

An authorized officer is in a position to enter a premises in order to silence an alarm if:

- The alarm has been sounding continuously for 20 minutes or for an hour on an intermittent basis
- The sounding of the alarm is likely to give persons living or working in the vicinity of the premises reasonable cause for annoyance
- If the premises are in an alarm notification area, that reasonable steps have been taken to get the nominated keyholder to silence the alarm

---

Under these circumstances the authorized officer can not enter a premises by force and must be able to show evidence of authority to act under the legislation.

A warrant to enter by force can be issued if a justice of the peace is satisfied that all conditions are met and that the officer is unable to gain entry to the premises without the use of force.

Before applying for a warrant the officer must leave a notice at the premises stating:

- That the officer is satisfied that the sounding of the alarm is likely to give persons living or working in the vicinity of the premises reasonable cause for annoyance, and
- That an application for a warrant, authorizing the officer to enter the premises is to be made, so as to gain entry to the premises using reasonable force if necessary for the purposes of silencing the alarm.

Silencing the alarm

The authorized officer is empowered to

Take any steps necessary for the purposes of silencing the alarm.
In doing so the officer can take as many other people and equipment as deemed necessary.

There must be no more damage caused to the premises than is needed to silence the alarm.

A notice is to be left stating what action has been taken.

Although there is no mandate to reset the alarm the premises must be as effectively secured against entry as how they were found.

Installers are obliged to familiarize themselves with the basics of the legislation so they are in a position to answer any customer queries.

There is an additional need to appreciate the time honoured Department of the Environment Directive on the control of nuisance noise as follows.

Department of the Environment Directive on the control of nuisance noise

How is the intruder alarm industry affected? The Noise & Statutory Nuisance Act 1993, Chapter 40, Schedule 3, states:

A person who installs an audible intruder alarm on or in any premises shall ensure –
(a) that the alarm complies with any prescribed requirements, and
(b) that the local authority is notified within 48 hours of the installation.
A person who without reasonable excuse contravenes this requirement or notification shall be guilty of an offence and liable on summary conviction of a fine.

This Act also makes provision to be made for expenses incurred by local authorities in abating or preventing the recurrence of a statutory nuisance to be a charge on the premises to which they relate; and for connected purposes.

This Act also empowers an officer of the local authority to enter a premises by force, if need be, to silence an alarm if the operation of it is such as to give persons living or working in the vicinity reasonable cause for annoyance. This must follow an application made by the officer and accepted by a Justice of the Peace.

The strength of the Act is clear. Governing codes which also relate and can be invoked are considered next:

● Code of Practice on Noise from Audible Intruder Alarms 1982
● Control of Pollution Act 1974. Part III – Noise
● Pollution Control & Local Government (Northern Ireland) Order 1978.

The first of these documents classifies noise made by the operation of audible intruder alarms as a frequent cause of complaint. This relates mainly to duration rather than to the volume of noise produced. Owners and occupiers who seek to protect their premises, whether private or commercial, by installing an audible system have a responsibility to ensure that a device fitted primarily for their own benefit does not become a nuisance to the public at large.

The code of practice promotes regular maintenance under a contract and the replacement of any unreliable or ineffective equipment. For identification purposes it asks that measures be taken to make it as easy as possible to identify the particular premises where an alarm is being generated. This is to help the police contact the keyholders as quickly as possible. One means is to adopt xenon strobe lights adjacent to the sounder to operate in conjunction with the sounder. This may continue after termination of the sounder output to indicate that the premises are still in alarm.

The code of practice requests that within 48 hours of installing a new alarm system or taking over an existing one the local police station be notified. This is to be done in writing, with the names, addresses and telephone numbers of at least two keyholders. These should be telephone subscribers, and have their own means of transport where necessary. They should be well versed in operating and silencing the alarm, and the alarm company itself can act as a nominated keyholder.

Section 4.1 of the code of practice relates to automatic cut-out devices. It states the desirability of fitting these devices, which stop the sounder after a period of 20 minutes from activation of the system. This 20 minute period is now time honoured in the intruder alarm industry.

Environmental Protection Act 1990, Section 80, is used by the local authority to serve notice of the existence of noise amounting to a nuisance by the excessive sounding of an audible intruder alarm. Under this procedure the noise is classed as a statutory offence and subject on conviction to a heavy fine.

Discussion points

With respect to noise there are a number of points worthy of thought and discussion.

- People may have resistance to an audible alarm being installed, and ask if any notice is taken of them and how it affects an intruder. The question in response is whether the burglar would be happier during an intrusion to create an upsetting noise or work in quiet?
- The importance of the correct secondary supply at the control panel and provided to the stand-alone self-activating bell (SAB) and self-contained bell (SCB) modules during a power cut is apparent.
- How is the requirement for audible cut-out devices satisfied by the manufacturers of intruder alarm systems and components?
- What extra measures must the installer take in the event that the hold-off voltage could be lost to SAB and SCB modules either accidentally or by some deliberate action?
- What precautions should be taken with SCB modules, often found in multiple sounder applications, when the alarm sounder is always supplied by the on-board battery?
- What difference can an open circuit make in the tamper loop? What fears do system purchasers have and how can these concerns be addressed?
- What must also be considered to ensure that a system cannot power a sounder to the programmed bell cut-off time but then persist in rearming and repeating the process of sporadic triggering until disarmed in total by the keyholder?

All customers are deeply concerned that an audible system may continue to sound for an extended period of time and these people need an assurance that the system caters for a fault that could create such a condition with the audible devices. The intruder alarm must incorporate the technology to negate such problems. The technical details of sounders and the role of the modules are given in Section 6.1.

1.7   The installation and servicing company

It is obviously not possible to be able to continue to trade unless a viable profit is made. We can only summarize and recognize that at the outset the product and service must be sold at the correct price.

Sell three things:

- yourself;
- your company;
- your product.

Yourself

The following aspects should be considered:

- Introduction – do so in a correct fashion. First impressions count: be punctual, courteous and helpful.
- Professionalism.
- Confidence.
- Appearance – clothing, grooming, personal hygiene.
- Presentation of documentation, literature, etc.
- Knowledge – security awareness; of crime prevention products, the competition, and the security industry as a whole; of intruder alarms, installation and maintenance; of interested parties, the police and insurers.
- Attitude.

Your company

The quality of presentation of the company will depend upon your personal knowledge of the organization. Ensure that attributes 1–6 apply:

(1) *Professionalism.* Understand:

- BS EN ISO 9000;
- response organizations;
- British Standards;
- IEE Wiring Regulations;
- ACPO unified police force policy.

Reiterate that membership of a trade organization or approval status proves a willingness to be vetted against a recognized standard.

(2) *Length of time trading*

- Young company: emphasize the ability to adopt new technologies without having to alter old patterns.
- Established company: show evidence of past successes and use the number of years trading as evidence of security.

(3) *Ownership*. Understand the meanings of:

- sole trader;
- limited company;
- subsidiary of a holding company;
- corporation.

Large organizations have more resources to support customer needs. Small firms can offer a personal service at a local level.

(4) *Competence*

- Training is to nationally recognized standards.
- Staff are suitably qualified.
- Personnel have experience through length of service.

(5) *Accessibility*

- Provide contact details of the sales office.
- Provide contact details of the installation and service department.
- Provide names of contacts and their telephone numbers.
- Provide 24 hour cover.
- Have a quick response time.
- Provide a good monitoring service.

(6) *Company policy*

- Good customer care and health and safety exemplify knowledge of and pride in the company.

Your product

Technical persons can be faulted in overdemonstrating their wisdom of the product by relaying technical details to the potential, usually non-technical customer, and this is generally not welcome.

Important points to note:

- do not use jargon;
- do not talk down to the customer;
- do not overelaborate as confusion may occur;
- listen carefully to questions;

- answer honestly;
- discuss your evaluation of the risk.

The customer will want to know:
- quality/reliability of intruder alarm systems;
- what exactly the product will do;
- what the product will not do;
- cost;
- suitability;
- installation technique – disruption, access to areas, number of people on site.

Within this there is a need for confidentiality.

Confidentiality

There is a need and a duty to protect information on customers and their security systems and to prevent the loss of this. Equally there is a need to harbour unrecorded information derived from visiting premises and conducting interviews with clients.

Criminals can acquire information in many ways, and customer information may be lost by a surveyor. Typical ways in which information can be lost and acquired by others are listed below:

- lost from car;
- lost from home;
- conversation with family and friends;
- conversation at work or through mobile telephones;
- indiscrete conversation with other parties during leisure;
- telephone conversations;
- documents being left unattended such as fax or computer data;
- waste paper not being correctly destroyed.

The surveyor is always working with confidential information, and must be alert to being 'pumped' for security details. Although salesmen must communicate well they must be careful to whom they communicate and what they divulge.

To maintain confidentiality, adopt a specific approach by:

- not using or naming examples of recent surveys;
- insisting that information will only be divulged when agreed by both the customer and your manager;
- recognizing the intelligence and determination of the criminal fraternity;
- not unwittingly supplying information to your clients' competitors;

- cross-shredding or burning all waste paper;
- never giving access to others in order that they may see customer details or specifications – this includes visitors to the office or cleaners;
- keeping files and other confidential material locked away;
- ensuring that a fax can only be received by the addressee.

Computer security

The main risk areas are:

- information left on screens for unauthorized persons to view;
- programs left running allowing access to information.

Disks and tapes must be protected. Security codes should be used for accessing data and for programs.

Information

In the main, information about the risk and the measures being considered to minimize the risk should only be given to those that have the right to know and then only that which is necessary to know.

At the office

Check enquiries by telephone first with the customer, and check the accuracy of the number and name. Provide only authorized information.

Do not leave information on printers and copiers.

Danger zones

- Mobile telephones: conversations may be overheard. Do not discuss names and addresses in connection with privileged information.
- Fax machines. Ensure that the intended recipient is available to receive confidential data.
- Dictation tape recorders. These should be kept secure. Tapes should be wiped after being transcribed.
- Portable note pad. Access codes should be used to protect data.

Summary

Maintain your own high standards of confidentiality and be secure in the knowledge that you will never be prepared to let your client, company or yourself suffer from even a temporary lapse.

2 Intruder alarm system circuitry

This chapter looks at the three key circuits used in an intruder alarm system, i.e., detector circuits, power circuits and control circuits.

It is important to be able to recognize the uses and limitations of the wiring methods that are employed in linking the detection devices with the control equipment. This can be achieved in many different ways, be it open or closed circuit and single pole, double pole or using two- or four-wire circuits. In addition, monitoring devices must be understood: these may take the form of diodes, end-of-line resistors or resistive networks and multiplex buses. Line monitoring is being increasingly used, even in lower-risk applications, and represents the future of detection circuits because of its labour-saving and reduced wiring potential. The self-latching relay continues to have widespread use in the integration of circuits where electronic switching and semiconductor technology is not available, so its application must be explored.

Power circuits are those that energize detectors and signalling equipment. They are governed not only by the equipment but also by the cable run and the considerations of supply voltage over long distances. Control circuits are employed to collect data by a number of methods including addressable and polling techniques and by multiplex configurations. Control circuits are additionally used to initiate functions such as walk test and latch and control plus remote detector reset. Control circuits need to be protected against damage, either accidental or malicious, and have adequate protection when in the proximity of electromagnetic sources.

The student must also realize the need to employ anti-tamper methods and how to specify remote safe shunts for detection devices. An understanding must also be gained of the methods of addressing items on the control circuit wiring network.

2.1 Detector circuits

Under the European Standards there are no restrictions as to how the detector circuits and the interconnection of components can be made. Therefore it can be by cable, wireless or any other technique providing that it satisfies the required level of monitoring for the appropriate system grade. Although cables still make up the majority of detector circuits there is now a greater move towards systems that use a variety of detection circuits with a mix of techniques. The different system grades dictate how the detector circuits must be configured and this is shown in Table 2.1. Grade 1 is DIY/mass consumer market only so single pole is generally acceptable.

Table 2.1   EN system wiring detector grades

Wiring type	Technique	Advantages/disadvantages
Double pole 4 wire	2 closed loops (circuit conductors) at different polarities. Loops provide supervision for 'detection' and 'tamper' by responding to a cut cable at all material times (24 hrs) **EN Grade 2 systems.**	Traditional and easily understood. All systems using this configuration can be easily retrofitted. Individual circuits are used but this reflects in greater cabling run lengths.
End of line (EOL) 2 wire	An EOL device used for the detection loop at the 'far' end of the circuit. The control equipment analyses the impedance.	Understandable and a progression on double pole. Manufacturers use different EOL values. Can not be retrofitted without amending the wiring at the final device.
Dual operation Fully supervised loop (FSL) or EOL 2 wire	Resistors are fitted in parallel in the detection device across the contacts. May also use series resistors plus an EOL **EN Grade 2 systems which require remote maintenance to be carried out.**	Provides full supervision with a reduction in cabling. Not easily retrofitted. Resistors are supplied by the manufacturer to suit the equipment type.
Triple resistors FSL 2 wire	Use variant value resistors for alarm contact, fault contact and EOL. **EN Grade 3 systems.**	Provides full supervision and fault contact anti-masking with circuit analysis of all wiring states. Not easily retrofitted.
Multiplex. Data bus. Addressable	Use main data highways, branches or node type networks (point in a circuit at which the conductors join to form a junction).	No requirement to wire each device as a 'home run' back to the main control point. Equipment differs enormously in terms of parameters and system configurations.

Open circuit

This may also be termed normally open (NO). It is effectively an open path in its free condition, and no current can flow. A switch or detector classed as normally open will have its contacts held open, and can only allow current to flow through it and then through the cables in the wiring system when it is either manually or automatically influenced so that its contacts are closed. This method is unsupervised and insecure because although the sensor monitors an area or a material it is not possible to monitor the cable for tampering or a fault.

Pressure mats, although rarely used now, are normally open devices, and alongside the normally open loop they carry a closed tamper loop. However, in the first instance we must understand the single-loop configuration as shown in Figure 2.1.

In Figure 2.2 one of the sensors has gone into alarm, but because the wiring is disconnected, no signal can be sent to the control panel. For this reason these wiring circuits are now rarely used, but the use of pressure mats should still be understood.

However, if the aforementioned circuit is used in configuration with a closed tamper loop, thus making up a double-loop or double-pole system

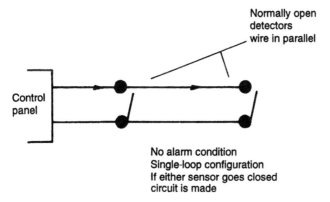

Figure 2.1   *NO single-loop configuration*

Figure 2.2   *NO circuit disconnected*

with the loops at different polarities, the control panel can respond to either *total* disconnection of the wiring or shortening of the loops (Figure 2.3). Normally open devices go across the loops.

Closed circuit

This may also be termed normally closed (NC). It is effectively a closed path in its free condition, allowing current to flow. The detectors in this circuit type are wired as a series that will go open when manually or automatically influenced, and this stops the supply current that would normally be flowing through the circuit. This type of detector circuit is capable of being supervised because disconnecting the circuit creates the desired open circuit. It is the chief wiring practice employed, but has certain disadvantages when only made up of a single loop or single pole (Figure 2.4). Clearly the wiring is monitored in this state for disconnection; however, if a short is placed across the loop either by a simple fault occurring in

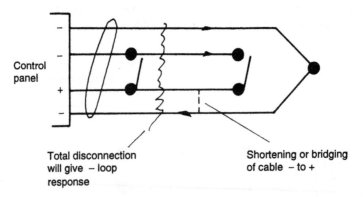

Figure 2.3   *Double-pole NO detector circuit*

Figure 2.4   *Non-alarm condition of a single-pole NC circuit*

the wiring or by an intruder shortening the detection loop, then the sensors could be activated without any control panel response being made (Figure 2.5).

Single pole

The single-pole system, although having some supervision, cannot cater for all events unless it is either combined with a further circuit or a pole with a different polarity or has some form of line-monitoring device. The normally closed circuits we have looked at have been single pole and also cannot give an indication of single-sided ground faults; additionally, double-sided ground faults can make the detector inoperative.

Double pole

It is normal for a detector to the ENs to be monitored on a constant basis for an open or short circuit, generally termed '24 hour' monitoring. It is also normal that an audible local tamper alarm be raised within 20 seconds and that an indicator shows the alarm condition. These indicators are to be on each processor and sensor so as to indicate the affected component. In a typical intruder alarm detection circuit, monitoring is performed using closed loops of different polarity for alarm and tamper, with the control panel giving a selected response for the different loops and also for different conditions of being set or unset (Figure 2.6). This set-up offers acceptable levels of security in most situations and will also cause an alarm if there is an earth fault on both sides of the loop. A single-sided ground fault would, however, not cause an alarm. The advantages of this wiring type are that the wiring and installation is easy and flexible, and low voltages and currents are employed. The disadvantage lies in the fact that such circuits could potentially be circumvented by an intruder with knowledge of the colour code being employed and shortening the positive loop. In practice this is not the

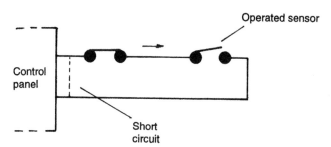

Even in the event a sensor has operated the short circuit inhibits the control panel from recognizing the activation

Figure 2.5 *Short across a single-pole NC circuit*

Sensors wired double pole N/C in four core cable
Shown in non-alarm condition
Tamper condition if open occurs in – at all material times
Alarm occurs if + goes open in set or + is bridged to –

Figure 2.6 *Double-pole NC detector circuit*

case in low- to mid-security applications, but in higher-risk situations line monitoring devices should be used.

Before looking at line monitoring or end-of-line devices we can add that the wiring of the detector to the alarm loop is a straightforward procedure. The switching contacts on a switch are easily established, and on powered detection devices the appropriate contacts are identified as alarm and tamper 24 hour.

2.2 Line-monitoring devices

So far we have seen how a conventional detection circuit can monitor detectors when they are configured normally closed. However, the only way in which a circuit can be totally supervised is by the use of line-monitoring techniques. With this principle, circuits are of the normally closed variety but with the current flowing in the circuit constantly measured and analysed by devices within the control panel, which must always detect an impedance within the circuit. An intruder may be prepared to attempt to short out a circuit but then must also influence this impedance, which is not easily done. Any change in current magnitude or direction will create an alarm condition even if the cable is not cut or it is shorted.

In the first instance consider a diode which will allow current to pass in one direction only. In the circuit described, it is used in conjunction with a terminating resistor (Figures 2.7 and 2.8). Observation will show that the diode will only allow the passage of current through the switches in one

Diode only passes current when this
end more positive than the other

Symbol

— light coloured
band

Silicon rectifier or equivalent

Figure 2.7   *Diode*

Polarity
reverse
in night

Terminating
resistor

Figure 2.8   *Polarity reversal*

direction. When 'day' is selected, the current can still flow through the circuit because the continuity diodes will short out the switches as they randomly open. However, in 'night' mode the polarity is reversed, and the diodes inhibit the passage of current when a switch opens. This contact opening then generates an alarm by signalling the control panel or releasing a relay. The terminating resistor is installed at the end of the line and its impedance must always be recognized by the control panel. If the circuit is shorted, the loss in impedance is detected. The resistor must always be fitted to the end of a line detector to ensure that the correct level of supervision is achieved.

Let us now look at a number of different ways of using an end-of-line resistor method.

Figure 2.9a shows the familiar standard circuit using a positive alarm loop with the switches or detectors wired in series, with the tamper loop as a separate pair, wired as a negative loop running within the same four-core cable.

Figure 2.9b is similar in style but employs an end-of-line (EOL) resistor at the end of the positive alarm loop. The control panel can then analyse the impedance and determine if the circuit is open and in alarm, if it is normal with the full impedance, or if the loop has been shorted and the impedance is no longer capable of being recognized. The advantage of this system is in its simplicity: it is not complex to those familiar with the standard wiring methods using four-core cable with separate alarm and tamper loops. It has the disadvantage that four-core cable must be used, although two-core cable can now be used to achieve the same effect by using one of two

Figure 2.9 *(a) Double-pole circuit; (b) Double-pole and resistor (EOL) circuit*

Figure 2.10 *Dual operation*

methods of dual operation (Figure 2.10). The principle of dual operation using two-core cable is as follows:

All devices closed	Loop resistance 4K7 Ω
Tamper open	Loop open circuit
Alarm contact open (EOL resistor + parallel resistor)	Loop resistance 9K4 Ω

In this example all devices are wired as normal with a 4K7 Ω resistor being fitted in parallel with the complete loop. The tamper is also wired as normal but with the 4K7 Ω resistor being fitted in series. The advantage of dual operation is that the same supervision as the familiar standard four-core cable operation can be achieved but with a reduction in cabling.

The next example is also a dual-purpose loop with an end-of-line resistor, but this type can be used with both normally closed and normally open detectors. It will be noted that the end-of-line resistor has a value of 2K2 Ω and is at the end of the loop. Higher-value resistors are employed across each pair of series contacts (Figure 2.11).

In the 'day' condition a continuous signal will pass through the circuit irrespective of the operation of the detection devices, although it will vary

Figure 2.11 *Dual-purpose loop with variant-value resistors*

as the detectors are tripped. It can be seen that it is also monitored for tamper open-circuit disconnection, so it follows that the circuit should never be open.

In 'night' set, if any detector is tripped the loop resistance will be affected and the current will drop, which will in turn cause an alarm to be generated by the control panel.

The system will also respond to a short circuit, effectively similar to the normally open device going closed.

Observations

In all cases the monitoring devices are placed within the detector. From this we can conclude that the detector circuit of Figure 2.9b is ideal for those wanting to use supervision and are already familiar with double-pole wiring.

Dual operation, as in Figure 2.10, when used with individual powered detectors is uncomplicated, but needs to be wired with care when a number of door contacts, or such appear on a single-zone loop.

The detector circuit shown in Figure 2.11 can be of great value if both normally open and normally closed detectors are to be supervised using two-core cable.

As previously mentioned the different EN system grades dictate how the detector circuits must be configured. This is illustrated in Table 2.1. The manufacturers of the equipment will claim an appropriate security grade for their equipment and the type of wiring to be employed. It will be found that there is an overlap of technologies between the different sources and the configuration of the control panel. However, Table 2.1 is very much representative of the intruder alarm industry practices in existence.

It will be found that with EOL and FSL techniques control equipment from different manufacturing sources will use different detector circuit resistor values. However, the installation manuals will include details of the specific configurations being employed.

Figure 2.12   *Three resistor wiring – Option M (Always Masking)*

Figure 2.13   *Three resistor wiring – Option F (Fault or Masking)*

Due to an increased use of masking detectors these can be considered in their own right. Two examples of the actual connections applied when using masking detection are shown in Figures 2.12 and 2.13.

Note: The term "input" in relation to the contact option means a circuit connected to a detector (the term "zone" is not used to avoid confusion).

The input has an attribute to inform the control panel that the third resistor is fitted. The choice between Option M and Option F is made by software configuration. The status is as follows:

Circuit resistance (Nominal) Ohms	Open contacts	Option M (Contact = Masking)	Option F (Contact = Fault)
SC	(Short circuit)	*Tamper*	*Tamper*
2K2	All closed	Input OK	Input OK
4K4	Extra contact	Masking detected	Fault
6K9	Alarm contact	Alarm	Alarm
9K1	Alarm + extra contacts	Masking detected	Masking detected
OC	Tamper contact	*Tamper*	*Tamper*

Figure 2.14 illustrates an example of a terminal connection configuration at the detector.

+12V	−	Walk	T	T	N/C	Masking detect. conn.

Power

To CIE

(panel)

2K2   4K7   2K2

Figure 2.14   *Detector terminal configuration*

It is important to also refer to Section 5.5 as this contains details of intelligent device (iD) wiring and other technologically advanced systems and their role in intruder alarm control equipment.

2.3   The electromagnetic relay

Throughout this book we will become aware of the use of the electromagnetic relay, and in Section 6.3 we describe its construction and how it can be applied as an interface.

In the present chapter dealing with system circuitry we must look first at the latching relay because it forms what is the original base security system circuit.

A simple switch-operated alarm can comprise normally open contacts and a latching relay circuit (Figure 2.15). In this case, when any of the switches closes, power is fed to the coil of the relay and the contacts close. The relay will then latch and stay latched even in the event that a switch opens, as the coil will be held by the relay contacts.

Observation

There are problems with the normally open circuit: any break renders the circuit inoperative, and twin leads must also run to each switch. Using a normally closed technique, any break in the wiring can be used to sound an alarm and then show a tamper or failure in the wiring. Single wires need only be run from the alarm to one side of the first switch as the next side of the switch is then connected by a single wire to the next switch in the chain, and so on.

In practice, Security Systems cater for both normally open and normally closed circuits. The base circuit is shown in Figure 2.16.

Consider the essential circuit. This is based on a CMOS two-input NOR gate, of which the 4001 integrated circuit (IC1) contains four.

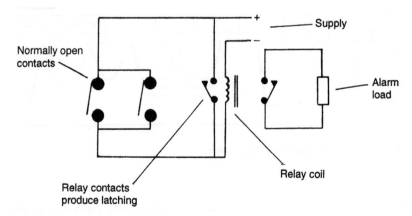

Figure 2.15   *Latching relay circuit*

Figure 2.16   *NO and NC security circuit*

The six inputs of the other three input gates are tied to the negative supply rail. For the two-input gate the output is high if both inputs are low and low also for the other three possible input combinations. The two inputs are configured in parallel so that the gate acts as an inverter. The inverter is taken to the high state by R1 and the normally closed switches. Under quiescent conditions the output of the circuit is low as the input is high.

In operation either a normally closed switch going open momentarily or a normally open switch closing will connect the inverter input to the negative supply rail. The relay coil will be powered, its contacts close and the alarm generated. It will only be possible to silence the alarm by switching

S1. This de-energizes the relay and causes the silicon-controlled switch IC2 to switch off. With the input switches set to the quiescent state, when S1 is again closed, the circuit is re-established.

Design considerations are in the use of protective diodes to suppress the high reverse voltage which would otherwise be generated across the relay as it de-energized. Resistors and capacitors are used to form a filter to remove noise spikes caused in the switch wiring which would cause spurious triggering. These noise spikes can easily be caused by the operation of electrical appliances or lightning.

This leads the engineer to appreciate how simple detector circuits, either normally open or normally closed, can be used to latch a simple system and how filters can be employed by means of resistors and capacitors. We should know how the detection wiring can be integrated into a simple network before considering power circuits and control circuit wiring, which are discussed next.

2.4 Power circuits

All powered detectors require a stable regulated supply or will go into alarm via the alarm relay loop if the voltage drops below a given value. Within the scope of powered detectors we class:

- movement detectors;
- glass break detectors;
- beam interruption detectors;
- vibration detectors;
- capacitive proximity detectors;
- acoustic detectors.

A number of years ago we would have quoted an operating range of the order of 10.5–15 V; however, we now see detectors claiming to operate at 6–17 V, with some functioning at even lower voltages such as 5.5 V.

Passive infrared detectors (PIRs) will essentially only draw some 20 mA current, and even combined PIR/microwave devices only take a 35 mA load. The beam interruption device will generally draw the greatest level of power, but this will still only be of the order of 50 mA when used indoors, although an additional 250 mA is required when heaters are employed outdoors.

Despite this trend, some consideration must be given to long detection runs and voltage loss over cable length. In domestic systems owing to the short distances, voltage drop is often negligible; however, in larger installations it is necessary to check that the drop in voltage from the supply terminals to the most distant load is kept within specified limits. This must also take into account the need for the secondary supply to be adequate.

We should recall that the electrical resistance, R, of a circuit element is the property it has of impeding the flow of electrical current and is defined by the equation $R = V/I$ where V is the potential difference across an element when I is the current through it. The unit of resistance is the ohm (Ω) when the potential difference is in volts and the current in amperes.

Consider a multicore cable consisting of 7/02 (seven strands of 0.2 mm diameter wire) and having a current rating of 1 A. The resistance is quoted as 8.2 Ω per 100 m. Hence, for a two-wire loop the total resistance is only 16.4 Ω. In practice this would have little effect on a detection loop as the current drawn is small.

If we now consider a PIR drawing 20 mA at 12 V the voltage drop only equates to 0.33 V, and as we know the nominal voltage is of the order of 13.7 V there would not be a problem unless multiple detectors were used on a long run; however, bad connections or strands of the cable lost in terminating will increase the voltage drop owing to increased resistance.

If we look at the infrared beam interruption device, used as a multiple of two, then the current can be 100 mA giving a drop of 1.6 V, which is reasonable. The heater will create a further drop of 4.15 V, drawing 250 mA over 100 m of cable. For this reason we should consider powering the heater through a separate power supply or via a heavier cable.

Table 2.2 shows how we can equate resistance per 100 m of twin cable and voltage loss over this 100 m when a current of 200 mA is derived.

In Section 5.5 we will consider multiplex systems which do not have all of the detectors wired in a direct fashion to the control panel or end station but

Table 2.2   *Voltage loss at 200 mA*

Wire size	Resistance/ 100 m (Ω)	Voltage drop at 200 mA (V)
1/0.1	438	87.6
1/0.2	115	23.0
1/0.6	12.8	2.56
1/0.8	7.2	1.44
7/0.2	16.4	3.28
13/0.2	8.8	1.76
16/0.2	7.2	1.48
24/0.2	4.8	0.96
32/0.2	3.6	0.72
30/0.25	2.4	0.48
40/0.2	2.9	0.58
0.5 mm^2	7.2	1.48
1.0 mm^2	3.6	0.72
1.5 mm^2	2.4	0.48
2.5 mm^2	1.4	0.28

use a data bus network with addressable points along the line. It will be found that strobes, bells, sirens and such are not powered from the multiplex cable but from local power supply units. The supply unit provides the 12 V power, and the multiplex module provides the trigger through an on-board relay. Consideration of Figure 2.17 helps us to recognize the method of using a dual detection loop, with a local power supply as a means of driving sounders and avoiding voltage loss. Detection loop uses 1kΩ resistors. In this case the power supply also powers the multiplex module.

We may conclude this subject by saying that Skills for Security believes that electrical fundamentals and basic electrical physics are important units in its syllabus because they underpin every other section of the learning programme. For example, using the mathematical term $V = IR$ the current can be predicted in a circuit when we know the terminal voltage of a control panel and the combined circuit resistance of both positive and negative poles is also known. If the terminal voltage is, say, 13.8 V and the combined circuit resistance is 66 Ω , then $I = 13.8/66 = 0.209$ A or 209 mA. This calculation can then be verified in the field using a test instrument, to ensure that there is no fault, such as leakage to earth (most control panels are referenced to earth). In practice, when faced with voltage drop considerations, most installers double up the supply cores to reduce the resistance of the supply conductors. We know that the greater the distance, or the smaller the cable, the greater is the resistance, and as the voltage at the terminals of the supply is fixed, voltage must be lost over the cable run. Damage to cables and poor joints themselves lead to a rise in the resistance of the supply conductors and a reduction at the head voltage or furthest point in the circuit. This head voltage is important, and is the terminal voltage less the sum of the voltage drops in the circuit. It can be both

Figure 2.17 *Multiplex module dual end-of-line resistor loop*

calculated and measured, and the results compared. Alongside this the student must be aware that there is an operating range for every device he or she is to install, and the effects of exceeding these ranges need to be understood.

2.5   Control circuits

We have learnt that the only way we may monitor circuits efficiently is by closed-circuit wiring, and with detection devices this is better when supervised by appropriate means. The detector wiring can then recognize any change in its state, and the control panel may then respond as appropriate, depending on whether the system is set or unset, and how it has been programmed to generate its output. Although the cabling is supervised, it is still necessary to consider how it may be protected from damage, either accidental or malicious, and we study this later in a survey of modern wiring methods (see Section 7.1).

Screened cable is not normally used unless multiplex networks are run through the same area as cables that produce radio frequency (RF) or are used to switch high current loads (e.g. mains supplies, telephone or computer lines). In any event, it is necessary to keep the network and detection cables remote from mains supplies, telephone cables, RF cables and those supplying bells or sounders.

With conventional wiring methods in which all the detectors are cabled back to the end station, the terminating of the zones and the programming of the zone types and attributes is found to be relatively straightforward, but with multiplex methods these detector signals are carried on a data bus network. It becomes necessary, therefore, to address the multiplex module before the programming at the main circuit board can be carried out. It may also be required to address the remote keypads mounted at various points in the protected premises. This is necessary so that the specific point of any setting or unsetting procedure can be logged. DIL switches are often adopted for this purpose. We will initially look at how the main control panel can recognize an individual keypad by its address code.

A unique binary address is given to each remote keypad by setting a four-way DIL switch, in the example of Figure 2.18. With the switch in the off position, it has a value of zero. With the switch on, it has a value of 1 2 4 8, reading from right to left.

The next example (Figure 2.19) shows a bank of eight individual switches numbered 1–8. In the on position they each count zero. In the off position they represent a different number.

Although DIL switches are the norm for addressing by binary code, a process of cutting out links to remove components can be used to select parameters, or applying jumpers across pins may be the preferred technique in some systems.

Figure 2.18 *Binary address: 4 bank*

Figure 2.19 *Binary address: 8 bank*

The method of returning a system to the factory default values and parameters is usually achieved by shorting certain pins during power up following total depowering. The menu-driven software will then permit the engineer to programme all the system functions: the details will be found to vary enormously between systems, although the zone types and attributes will be defined the same.

An important control circuit of any panel is the latch. At present, this is mainly used for inertia-type sensors. The latch output turns on at the end of the exit time (system set) and then turns off at the beginning of the entry time (system unset). It also turns on at the beginning of a walk test.

The new generation of control panels, conventional and multiplex, is multizone, so detectors such as PIRs can have their own unique zone to meet the British Standards requirements. A common practice is to apply the latch terminal to the detector, which enables the LED on any detector that has been alarmed to be latched for identification purposes. Unfortunately, this means a greater use of cabling or one extra core in the cable.

If we refer to Chapter 5 and Figures 5.1 and 5.2, the wiring technique of using latch with freeze and walk can be realized, and the means of shunting as an ancillary practice can be understood by considering Figure 5.13.

3 Intruder alarm detection devices

This chapter is concerned with the component devices used in the industry to detect intrusion. Included in this section are references to BS 4737 since many component standards to the ENs remain incomplete. However, although the actual technologies used to detect intrusion have remained the same in the move towards the European Standards there have been certain upgrades in techniques and the way they are applied.

We know that the European Standards state that appropriate manufacturers' compliant equipment and components must be installed to the appropriate system grade that has been established in the risk assessment.

In section 1.3 covering security grading the minimum levels of supervision were covered and related to the extent and type of intruder detection devices and protection that is to be installed. This applies to the detection of doors or windows becoming opened plus any attempts at penetration through perimeters to include walls, ceiling or floors etc. These all have to be determined along with the space protection needed in response to the premises and the contents held within it.

Having determined the system grade and the detection devices to be employed it then becomes possible to specify the exact components to be used as these will be supported by the manufacturer as compliant with an appropriate grade.

The following is an example of an adhesive label applied to a roller shutter contact and illustrates how a manufacturer will tend to mark a simple compliant product:

This product is suitable for use in systems designed to comply with PD6662:2004 at Grade 2 and Environmental Class II

Suitable for use in EN50131 security Grade 2 systems

On the face of it intruder detectors are rather simple devices but it is difficult to produce units capable of reliably detecting intruders and totally resistant to false alarms. Modern technologies must be used to reconcile the key objectives and advise installers on how to select the most effective product. Allied to this as manufacturers strive to introduce innovative and efficient products as a means of minimizing false alarm rates in line with the EN regulations new products have been launched.

As we know EN50131 changed the security industry enormously. The insurance industry was always keen on reducing risk and it set Grade 3 as the normal risk or accepted standard for the majority of commercial applications. Part of the change related directly to internal detection devices as systems were required to be updated to enable better signalling and control panel event memory to log detection data. Anti-masking of detectors became a major issue since movement detectors at Grade 3 must be able to detect 'masking' i.e. recognizing that something has been placed over the lens of the sensor. This was intended to initiate a test procedure by the sensor element and to signal the control equipment that masking had occurred. In turn the condition could be notified by a local alarm or even at the ARC.

Anti-masking means that any attempt to mask a sensor by applying such things as sellotape, spray, glue or adhesive paper over the detector element would trigger a mask alarm. Dual-technology detectors need both technologies to work to confirm an alarm therefore any mask of any sensor element must be notified to the control equipment.

It is to be noted that ultrasonic technology, although rarely used now in isolation, has been included in this section because it is recognized under the ENs with passive infrared as a combination technique.

This chapter concerns itself with the range of detectors currently available. It considers them in terms of principles of operation, false alarm avoidance, set-up procedures and identification.

The different versions of protective switches, both mechanical and magnetic, will be described, including mercury tilt contacts, microswitches and the most common magnetic reed switches. Contact forms that will need to be understood include momentary break and changeover.

The application of continuous wiring either directly attached, or in tubes, together with transverse methods are included in this chapter, and so is foil on glass, although now replaced in many applications by piezoelectric and acoustic glass break detectors.

Deliberately operated devices for use in times of immediate threat, together with the specific demands imposed upon their installation, are investigated. Beam interruption detectors are studied alongside movement detectors and the technology that they employ, be it passive infrared, ultrasonic or microwave or a combination within one enclosure. Another powered detection device is the vibration detector, which may be directly connected to the control panel or have its own particular analyser. The role of capacitive detectors and acoustic detectors in the intruder industry must also be understood.

The student must become aware of the sensor activation energy source and operation of all of these detection devices.

The relative advantages and disadvantages must be appreciated in order to be able to select the most suitable unit for a given situation, but also with account taken of design features that must be incorporated in the device to reduce false alarm hazards. There are false alarm hazards that can apply to any detection device, and in the case of movement detectors it is vital not only to identify the best position for siting in a given position to maximize effective coverage but also to minimize false alarms. Knowledge of the setting up procedures is therefore vital.

It must be stressed that the student must appreciate the activation method of any sensor in order to judge its role and to select correct units. Reference is made to the governing standards for the detection devices as applicable, and any particular requirements are covered. It must be realized that the scope of detectors is wide-ranging, and so certain base types may have many variants whilst other sensor technologies are more specialized. In the event that it is possible to easily identify the most employed variants of a sensor type, its details such as activation method and advantages and disadvantages will be provided. The more diverse technologies will be described in overviews.

3.1   Protective switches

Of all the electronic and electrical sensors available for detection or for completing a circuit in order that an electrical current can flow through it, the switch was one of the earliest compounds devised.

It is used in high volume in the intruder alarm industry as a protective device and has some merits that surpass all other detectors.

The magnetic reed switch

The magnetic reed switch is the most commonly employed switch, and it exists in a wide variety of forms.

Essentially it comprises an internal mechanism of two slender metal reeds hermetically sealed in a glass tube and positioned so that their ends overlap slightly but do not touch. The contact material is gold plate, platinum or an alloy of rhodium or precious metals.

Actuation is by the placing of a magnet in close proximity, which causes the reeds to make physical and thus electrical contact by mutual attraction. When the magnet is withdrawn, the elasticity of the metal reeds causes them to spring back to their free position.

The reed assembly is embedded in a plastic case which may be further protected in a metallic enclosure for use in applications where mechanical impact could more easily occur. The operating gap is the maximum quoted distance which must be allowed between the surfaces of the enclosure of the reed switch and that of the attracting magnet. The detector is therefore of two parts: a permanent magnet which is installed on a moving door, window, roller shutter or such and a reed switch which is fixed to the frame.

The magnet is a simple two-pole hard-core permanent magnet with high life expectancy and reliability.

The characteristics of the magnetic reed switch are easily understood and can be summarized:

- *Types:* surface or flush mount; single, double or triple reed; balanced or coded.
- *Activation:* displacement between a moving part and a frame, usually a door/window/roller shutter/shutter; the value of the displacement varies but should be 100 mm maximum.
- *Variant uses:* cabinets/drawers.
- *Not recommended:* as a sole detector for high value goods.
- *Advantages:* easily understood; reliable; best used concealed.
- *Disadvantages:* movement of the magnet must occur, so the switch cannot respond to cutting through the window or other moving part.

The versions most commonly employed are:

- Quick fit
  Flush fitting. Five-screw rectangular fascia plate. Operating gap 20 mm. Cut-out 20 mm.
- Speed fit
  Flush fitting. Five-screw circular fascia plate. Operating gap 20 mm. Cut-out 20 mm.
- Miniature prewired
  Flush fitting. Four-wire sealed. Operating gap 15 mm. Cut-out 8 mm with fixing by pins.
- Miniature prewired
  Flush fitting. Four-wire sealed. Operating gap 15 mm. Cut-out 8 mm with fixing by wedge action.
- Surface contact
  Five screws. Operating gap 20 mm. Visible size 67 × 14 × 13 mm.
- Surface contact
  Prewired. Operating gap 20 mm. Visible size 67 × 14 × 13 mm.
- Surface contact
  Miniature prewired. Operating gap 16 mm. Visible size 38 × 9 × 16 mm.
- Patio contacts
  Five screws or prewired. Surface mounted. Visible size 48 × 18 × 10 mm. Aluminium anodized or white finish.
- Roller shutter contacts
  Prewired. Die-cast aluminium enclosure. Operating gap 50 mm. Wiring protected by a flexible stainless steel sleeve.
- Heavy duty
  Prewired. Die-cast aluminium enclosure.
- Angle contact
  Mounted on a bracket. Operating gap 28 mm.

All versions: voltage range 1–100 V; 500 mA switch (max.); contact 0.5 Ω; temperature range –40 to 50°C.

Standard: BS 4737: Part 3: Section 3.3: 1977, 'Requirements for detection devices. Protective switches'.

The reed switch can be seen as a device that provides a closed circuit when the magnet is presented to it. The switch and magnet are either mounted within the frame and movable opening (flush) or on the outside (surface). Reed switches are essentially simple and reliable, but there are certain deliberations to note:

- Use on good fitting openings not subject to excessive vibration.
- Ensure that the opening allows the reed to switch within 100 mm.
- Steelwork influences the magnet so non-magnetic interfaces must be applied between the steelwork and the switch and magnet. Thin wood is adequate.
- The wiring is best concealed. Simple surface switches can be defeated by introducing a further magnet.
- Install with the axis of the reeds parallel to the face of the magnet.
- The application of flush switches involving the drilling of frames can draw dampness into the frame.
- Customers may have resistance to the installing of any contacts for aesthetic reasons, particularly in the domestic sector.
- Use sealed prewired versions where dampness may be a problem.

The type of reed widely used is the single reed, and it is of great effect when supplemented by other detection devices with different techniques. However, for more specialized use or for specific high-security protective switch applications it is possible to source double- or triple-balanced coded reeds. These have changeover or multiple switching operations with auxiliary circuits to provide monitoring. They may include biasing magnets or they may be shielded from any magnetic field not induced by the installed magnet. The poles are sequenced so that an additional magnet cannot be used to cheat the reed switch. In the event that further magnets are applied to the reed, the system can give a programmed output.

In terms of EN50131 the current interpretation of PD6662: 2004 is that the magnetic reed switch must be graded as 1, 2, 3 or 4.

Grade 1 contacts do not need to be tampered. Grade 2 switches should be tampered or sealed (potted) and Grade 3 contacts are to be complete with factory fitted resistors unless they can be fitted at the installation stage.

Microswitches

A microswitch is an electromechanical device in that when a mechanical force is applied to its actuator its contacts will be moved. These actuators are available in many forms of plunger or lever. They are generally found as anti-tamper switches protecting enclosures and housings or forming the switching element in deliberately operated devices. The microswitch

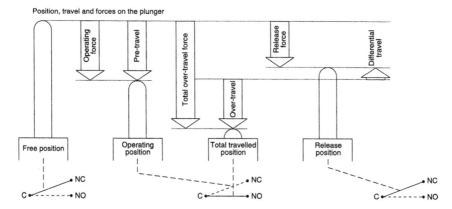

Figure 3.1 *Microswitch positions*

also has the capacity to handle heavy loads directly, and in the early days was used to carry loads when influenced by a door opening as in a simple circuit. The risk of false activations with the microswitch is virtually non-existent, and its characteristics differ from those of the reed switch (Figure 3.1). The contact layouts are also available in many forms. Figure 3.1 shows a changeover configuration switching one pole.

The microswitch still has a role to play, but in the main it is an auxiliary device to monitor enclosures against tamper, although it may actually be used as a detection unit if it can be made immune to being accessed in such applications as low-security monitoring of doors and other openings.

Mercury tilt contacts

The mercury tilt switch comprises an encapsulation that contains a pair of contacts bridged by a ball of mercury which is free to roll around the interior. The encapsulation is sealed and filled with an arc-suppressing inert gas. The on/off state is dependent on the attitude of the switch relative to gravity. It is sensitive to sideways and vertical displacement and will rapidly respond to sharp movements or a knock, but will remain closed for gentle movements.

This type of switch is mainly used as a tipover device to detect vibration or motion, and is often found on windows that tilt to open.

Variant types can be used normally open, but when vibrated the pulse may be used to set off an alarm, and is often found as such in mobile security units for vehicles.

It is also used in external applications when, for example, fence sensors are found to be influenced by the wind or high-frequency vibrations.

3.2   Continuous wiring

Continuous wiring was the original defence against the cutting of a large enough hole in a door or partition to permit physical entry. It is used in the form of taut wiring across an opening which if displaced to any great extent creates an alarm. In practice it can be applied in a number of different ways.

- *Types:* continuous wiring; wire in tubes; taut trip wire.
- *Activation:* distortion, cutting, bridging or breaking.
- *Variant uses:* openings or to prevent access through walls/partitions.
- *Not recommended:* in areas of high moisture.
- *Advantages:* easily understood; reliable.
- *Disadvantages:* labour intensive; not easily reset after intrusion.

Standard: BS 4737: Part 3: Section 3.1: 1977, 'Requirements for detection devices. Continuous wiring'.

The wiring is hard-drawn copper of 0.3–0.4 mm gauge. The insulation is PVC, and should be 0.2–0.3 mm thick. The fixing points should be no more than 600 mm apart, and the spacing between adjacent runs no more than 100 mm. If in tubes or grooved rods the spacing between adjacent runs should be the same, and they should be supported at not more than 1 m intervals. If recessed into supports the amount of recess should be no more than 5 mm but less than 10 mm from the end of the tube.

The wire emanating from the tube ends should be supported at no more than 50 mm from the tubes. Less than 50 mm of displacement should be necessary to break the wire and generate the alarm.

The wire itself is of low electrical resistance and is very brittle, so that it must be handled with care when attaching to any structure.

When applied as continuous wiring on walls, doors, partitions, screens or cases it is best fitted transversely, running vertically and horizontally in direction. The wires should be rigidly fixed at not more than the specified 100 mm intervals so that it is well supported. Even a small movement should be enough to stretch the wire. The wiring should then be concealed and covered with a panel for protection. Door loops can be used to bridge opening partitions or such.

When installed as wire in tubes these must be on the secure side of the protected premises. The tubes may be metal or hollow wooden dowels. Aluminium tubes are widely used for aesthetic purposes. The tubes should be vertically fixed to resemble window bars, and at a maximum distance apart of 100 mm. The wire is passed through the tubes and connected at the ends to a metal frame which holds the dowels. The tubes themselves should not be greater in length than 1 m.

A trip wire is applied as a single taut lace wire but rigidly fixed across a small area. The term 'trip' refers to a single wire. Trip wires are subject to the same requirements as the other methods.

Although continuous wiring as a category can also include wiring on glass, or wired glass where the wire is cast into the ordinary annealed glass, we consider this as being related to foil on glass, to which it is more relevant.

3.3 Glass break detectors

Glass break sensors are an added dimension to perimeter protection in that they monitor glass that is likely to be broken by an intruder to gain entry and can then generate an alarm with the intruder still outside. They can therefore be seen as helping to reduce property damage and confrontation by frightening the intruder before he or she gains entry.

Types of glass

The principal glass types that must first be understood are plate, tempered and laminated. The different glass types look similar when installed, and all three types may be found in the same building; for this reason the sensor must be capable of detecting the breakage of any type of glass without adjustment.

Plate glass

This is the type of glass most of us are familiar with. It is typically found in homes and older commercial buildings. When it breaks, plate glass shatters into large jagged pieces. It is not used in glass doors because of the risk of injury.

Tempered glass

Often referred to as safety glass, tempered glass is plate glass cut to a particular size and then fired in an oven at 750°C which changes the internal structure of the glass. When it breaks it shatters into thousands of tiny pieces to reduce the risk of injury. It is found in the majority of commercial windows. In the home it is used in glass doors, side windows next to doors, low windows and other areas where the risk of accidental breakage is high. Tempered glass is less convenient to work with than any other glass types because it cannot be cut to size at the job site and must be ordered in advance.

Laminated glass

This comprises two panes of plate glass laminated together with a plastic polymer film between them. It is extremely strong and difficult to break. When laminated glass is broken, the plastic inside absorbs much of the breaking energy and holds the glass shards together. It is commonly termed 'safety glass', and often found in commercial applications such as business offices and retail stores. It can be cut to size and is often used to replace broken tempered glass.

Characteristics of glass breaking

Shock frequencies are produced when glass breaks, and these travel through the glass and often through the surrounding walls and ceilings. The most intense frequencies are in the range of 3–5 kHz.

Acoustic frequencies are generated in a band ranging from the infrasonic (lower than we can hear) through the audio band to the ultrasonic level (higher than we can hear). This leads the engineer to the two main types of glass break detectors, namely piezoelectric and acoustic:

- *Piezoelectric detector:* senses the shock frequencies that travel through breaking glass and often the surrounding frames and walls.
- *Acoustic detector:* reacts to the sound frequencies associated with glass breaking.

Piezoelectric and shock sensor technology

As glass breaks, some shock frequencies travel through the glass while others migrate to the frame and still others into the walls and ceiling.

Glass-mounted sensors generally protect only one pane whilst those designed for installation on the frame can, in many instances, protect several panes.

Windows that open do not have a solid connection to the frame, and it should be noted that the less solid the fit the more difficult it is for the shock frequencies to travel off the glass. It is not possible for shock frequencies to reach frame-mounted sensors in windows with felt or rubber linings.

The new generation of shock sensors are electronic, using piezoelectric transducers to detect the shock. The piezoelectric crystal is tuned to the specific frequency of 5 kHz, which is always present when glass breaks. The tuned piezoelectric crystal is compressed by this frequency, and produces electricity which can be detected.

The sensor is a two-wire device with no power from the loop, and the only way it can alarm is if it generates its own electricity, so piezoelectronic sensors have false alarm immunity.

Installation

Shock energies concentrate in the corners of a sheet of glass, and the sensor should thus be mounted in one of these areas.

Manufacturer-quoted ranges for the detection distance are given to the point of impact, so if a point of impact is beyond that range, the sensor will miss that break even if the entire window breaks.

The installation of fixed-sensitivity shock sensors should be considered since they significantly reduce the installer's chance of error in setting sensitivities too low.

Acoustic sensor technology

There are three categories:

- single- and dual-frequency detection;
- dual technology;
- pattern recognition.

Single- and dual-frequency sensors

Single-technology acoustic sensors detect the sound of breaking glass at one or two frequencies. This is done by the use of a microphone with processing circuitry analysing the sounds to distinguish between glass break and other events. The sensor only has about one-tenth of a second to determine if the signal is real or a false alarm source.

The original acoustic sensors detected a high-amplitude (loud) sound at the frequency where glass break frequencies peak, typically 5 kHz (Figure 3.2). These were effective but susceptible to false alarms from other high-amplitude sounds at the same frequency.

Immunity to false alarms was improved by detecting loud sounds at two (dual-frequency detectors) or three characteristic frequencies (Figure 3.3). However, if the sound at one of the frequencies is not loud enough, no alarm can be generated: blinds and drapes reduce the loudness of any break, and an intruder may break glass quietly to avoid detection. The dual-frequency detectors therefore also had shortcomings.

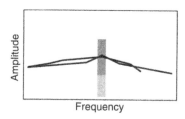

Figure 3.2 *Single-frequency acoustic sensors detect loud sounds at a single, predefined frequency*

Dual-technology sensors

Dual-technology sensors adopt two different technologies in one unit to determine an event. They must receive an airborne acoustic signal and at the same time detect mechanical vibrations travelling through the surrounding structure (Figure 3.4). These devices improve false alarm immunity but more emphasis must be placed on installation and testing to ensure that detection is not compromised.

Pattern recognition technology

This type of sensor provides better false alarm immunity than either the dual-technology or dual-frequency sensor. Rather than listening for a loud sound at one or two characteristic frequencies it encompasses much of the glass break frequency spectrum and examines the timing of the signal.

When the entire frequency spectrum is considered, sound patterns such as clanging pots and pans or slamming doors look very different from glass break patterns. Since the glass break pattern is present whether the glass is broken loudly or quietly, pattern recognition allows detection of the quiet breaks that the previous glass break sensors may miss.

Achieving false alarm immunity without compromising detection requires complex circuitry and comprehensive acoustic signal processing. To achieve

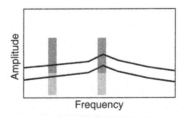

Figure 3.3   *Dual-frequency sensors detect two characteristic acoustic frequencies travelling through the air*

Figure 3.4   *Dual-technology sensors require a simultaneous acoustic signal and a mechanical vibration to generate an alarm condition*

this end, custom-designed application-specific integrated circuits (ASICs) are used. These replace the numerous conventional integrated circuits which would otherwise be needed. Unlike a microprocessor, which can only process data sequentially, an ASIC processes multiple data points simultaneously, and thus can process up to three times more data in the critical first 100 ms of a glass break event (Figures 3.5 and 3.6). The ASIC allows the sensor to rapidly process multiple parameters, such as signal duration, amplitude, frequency pattern and rate of change simultaneously.

To keep sensor size to a minimum, surface mount technology is used with careful placement of integrated circuits and radiofrequency filters to suppress radiofrequencies. Transient voltage suppressors provide suppression of electromagnetic spikes from near-strike lightning, compressors, HVAC motors and other sources causing power surges.

Installation

Acoustic glass break sensors cover an entire wall or room of glass. They can be mounted on the ceiling, on adjoining walls, on the wall opposite the glass or on the same wall as the glass. Microphones tend to be omnidirectional, so placement is not as critical as with older-generation sensors.

Acoustic sensors are designed to detect an impact that causes a hole big enough to reach a hand through. They cannot detect an impact by a bullet or an impact that only cracks the glass.

Acoustic detectors should always be mounted behind heavy drapes, never in front.

Figure 3.5 *Sensors with microprocessor technology process data one piece at a time or sequentially*

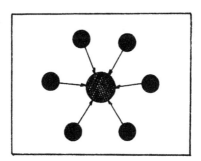

Figure 3.6 *ASIC technology processes multiple data points at the same time. The result is increased speed and accuracy during the critical first 100 ms of a glass break event*

Considerations

- For 24 hour loop protection use shock-type sensors. If protecting a single pane of glass, a shock sensor will be less expensive, provide excellent detection and be active 24 hours.
- For large walls or rooms of glass select an acoustic unit.
- If heavy drapes are present, mount devices behind them.
- Acoustic sensors can only operate in direct line of sight and cannot sense around corners or through furniture, etc.
- No sensor can detect bullet impacts or impacts only creating cracks.
- Acoustic detectors detect glass break frequencies and not those related to plastics.
- Never install beyond the specified range.
- Avoid installing acoustic sensors in small, noisy rooms.
- Avoid installing acoustic sensors in rooms with white noise such as that emitted from air compressors, or static-like sounds, as these saturate the sound frequency spectrum and can cause false alarms.
- Room acoustics must be considered during installation: an acoustically 'hard' room with bare floors and scant furnishings can become a 'soft' room with the addition of carpets, furniture and blinds – it is not then so acoustically live.

The acoustic sensor can be mounted temporarily in an intended position until it has been tested, which can be aided by means of a glass break simulator. These simulators are specific to a particular manufacturer's detector, and using other simulators can give an inaccurate indication of range. It should be remembered not to mount these acoustic sensors outside, or in rooms with high-level noise sources. The presence of air compressors, bells and power tools is troublesome. As an example of testing we can consider a device that has a red LED to indicate full alarm, such as when both flex and audio-signals are correctly received, and a green LED to indicate only that non-alarm noise signals are being captured. Table 3.1 summarizes how these LEDs typically operate when enabled.

Table 3.1 *Glass break LED response*

Condition	Green LED	Red LED
Normal, no event	OFF	OFF
Normal, event detected	Flicker	OFF
Normal, break detected	OFF	ON
Power-up self test	ON, one second	ON, one second
Trouble detected	Flash ON/OFF	Flash OFF/ON
Test mode, no alarm	Flash once per second	OFF
Test mode, event detected	Flicker	OFF
Test mode, alarm	Flash once per second	ON

Foil on glass

This detection method has now largely been overtaken by glass break piezoelectric and acoustic devices, but nevertheless should still be understood.

The detector comprises a foil of an alloy of lead and tin, or aluminium, or lead, with a self-adhesive back. It is designed to be a good conductor of electricity but have a low tensile strength, so that it breaks easily when the glass it protects is broken.

- *Types:* wired on 24 hour loop; single or double pole with alarm loop also.
- *Activation:* breaking or bridging.
- *Variant uses:* plate or tempered glass only (mainly for small retail commercial premises).
- *Not recommended:* for laminated glass.
- *Advantages:* simple and reliable.
- *Disadvantages:* easily damaged; not quickly repaired; poor aesthetics.

Standard: BS 4737: Part 3: Section 3.2: 1977, 'Requirements for detection devices. Foil on glass'.

The foil should be less than 0.04 mm thick and less than 12.5 mm wide. If used in the single-pole mode it can be applied as a rectangle between 50 and 100 mm from the edge of the glass. It may be run as a single strip not less than 300 mm long if the short dimension of the window is less than 600 mm. In this case it would be run through the centre, parallel to the long edges. It can also be laid as parallel strips not more than 200 mm apart, the ends terminating 50–100 mm from the edge. Unframed glass can be protected by a loop with dimensions of not less than 200 × 200 mm.

The foil is laid on the inside surface of the glass in strips which are generally available in a width of 3, 6 or 8 mm.

Installation

- Ensure that the glass is dry. Remove grease using methylated spirits.
- Take time to carefully apply the foil. Corners are a problem – do not attempt to join the foil – first fold the tape in the opposite direction, making a crease of 45° and then fold it back in the required direction. Do not run the foil over cracks or butt joints in unframed glass.
- The main run should be taken to a 24 hour loop, and the extra pole to an alarm zone for double-pole protection.
- Adjacent runs should be no more than 200 mm apart, and runs close to the edge of the glass are not secure so coverage of central areas is essential.
- Terminal blocks are fixed to the glass at the side of the pane. These make contact with the foil strip with terminals for interconnecting the wiring. 'Make-off' strips enable a bridge to be made across two panes.

- On windows that open, connecting blocks should be fitted to the window on the hinge side, and flexible leads bridged over to another block on the frame to connect the circuit wiring.
- Mount connectors at the top of the window to avoid problems with condensation or damage by objects placed on the sill.
- Connectors should be within 100 mm of the frame.
- A clear varnish can be applied to protect against corrosion once the installation is complete.

Examples of foil layouts are given in Figure 3.7.

All dimensions in mm

Figure 3.7   *Foil layouts*

3.4   Deliberately operated hold up devices

These are the devices which we know as panic buttons or personal attack or panic alarms (PAs). These are classed as hold up (HU) alarm triggering devices. They are governed by BS 4737: Part 3: Section 3.14: 1986, 'Specification for components. Specification for deliberately-operated devices'. We can consider them as comprising three types:

(1) those requiring a single force on one element;
(2) those needing two simultaneous forces on two different elements;
(3) those operated by two consecutive forces applied to different elements, the first being maintained while the second is applied.

Deliberately operated devices have the following attributes:

- Sensors may be latching or non-latching.
- For a manually operated device, the force required for operation should be within 4–5 N as applied by a 6 mm diameter disc.
- For pedal-operated devices the force should be within 5–8 N applied by a 12 mm diameter disc.
- Operation is by pressing a button or buttons on the unit if the user is threatened or needs attention. They must be wired to generate an alarm output at all material times.
- Their location must suit the premises. In the domestic sector they are found adjacent to the front door and at the bedhead in the main bedroom. They may also be used to great effect close to counters and tills in shops, banks and retail outlets.

Types 1 and 2 are available as:

- *Single front-push latching.* Operation by reed or microswitch. Reset by a key.
- *Single end-push latching.* Operation by reed or microswitch. Reset by a key.
- *Single front-push non-latching.* Operation by reed or microswitch. Must have control panel indication.
- *Single end-push non-latching.* Operation by reed or microswitch. Must have control panel indication.
- *Double front- or end-push latching.* Operation by reed or microswitch. Reset by a key. Ensures that both buttons are pushed simultaneously to give greater confirmation of wanted activation.

All the variants are normally available with cases made of ABS plastic or brass, and in sizes of the order of 80 × 65 × 26 mm. Other specifications are:

Contact ratings	0.5 Ω
Maximum switching current	500 mA reed or 10 A microswitch
Voltage range	1–100 V (reed) or 250 V (microswitch)
Contact material	rhodium (reed) or silver (microswitch)

Type 3 typically has two buttons, often behind a tactile keypad, and a preset with pretimed action capable of operating in a number of phases. Pressing one of the buttons will set the device into an automatic time delay mode. Failure to press the other button within a preset time will cause the device to activate. Pressing the button within the time limit cancels the timer and restores the unit to standby – an obvious use is that on going to answer a call the operator presses one of the buttons to set the automatic mode but if he or she is unable to return, the alarm signal is generated.

The device types mentioned so far have been those units which are most frequently used and that are simple in operation. They will be wired on a dedicated PA loop and have a unique channel when connected to remote signalling equipment, and in such cases may be programmed at the control panel to give no local audible output but to provide remote transmission only.

These devices must be capable of operating after temperatures of -1 to $+55°C$ have been applied to them for 16 hours, and able to withstand impact blows of 1.8–2.0 J. They must also have a working life of up to 10 years, and still be capable of operating efficiently even though unused for 12 months. An operated indicator must also be incorporated in the design. This is generally in the form of a small window that will change colour once the mechanism has been switched. The PA is further classified by the noise it generates during operation: normal (60 dB(A)) or quiet (30 dB(A)).

Hand-held or pendant-worn portable PAs can be used with hardwired systems by interfacing the radio receiver with the control panel PA loop.

Kick bars are a variant device which are floor-mounted steel arch enclosures containing the switch element. As with other versions, they must be able to confirm their operation. They are located at the specific point where the user's foot can access the kick bar easily and quickly.

We may conclude that PAs are intended for use in circumstances of immediate danger or physical threat to the end user. The knowledge that the user has ready access to a device which can bring immediate attention to his or her area or person provides a greatly increased sense of physical security. For situations which demand an increased sense of this security, the PA offers definite benefits.

It is important to note the requirements of Grade 3 systems in relation to panic buttons. If a PA is to be used as part of a Grade 3 installation, not only must it be tampered, but in the event of the cover being removed only a tamper signal should be generated and not the 'hold up' or 'PA signal'.

It is also important that communicator signals are selected correctly. This is because triggers can also be generated direct from keypads and inputs in addition to direct from a stand-alone PA device. In general:

HU device   triggers only for an alarm generated from a **HU input**.
Duress   triggers only for an alarm generated at a **keypad**.
Hold up   triggers for alarms generated by an **input** or at a **keypad**.

3.5 Beam interruption detectors

Infrared

These were introduced in the 1930s and are still in regular use today. The detector comprises two main components, namely an infrared (IR) light transmitter and a receiver. When an intruder interrupts the signal between the transmitter and receiver the alarm output becomes energized. IR photoelectric devices operate at wavelengths in the region of 900 nm at a carrier frequency of 500 Hz. In early sensors the IR beam was generated from a bulb with a glass filter which only allowed IR frequencies to pass. To pulse the beam a metal disc driven by an electric motor revolved in front of the bulb. Modern devices use an IR LED, within a protective enclosure, which is filtered to operate at the correct frequency. It is pulsed rapidly to give a concentrated beam of IR energy that does not generate heat. The receivers are contained within a single chip and are highly efficient: they recognize and capture the IR light energy and use a photoelectric cell to transduce the energy to hold the alarm circuit in a quiescent condition. The IR sensor in this case is known as an active device because it actually transmits IR energy. It is not to be confused with a passive IR device, which is passive in that it only recognizes the presence of IR energy at a given level. The IR beam interruption device is covered by BS 4737: Part 3: Section 3.12: 1978, 'Specification for components. Beam interruption detectors'. This standard states that an alarm should not be generated if a beam is reduced in intensity by less than 50 per cent. It also states that the alarm signal is to be in excess of 800 ms in length for an interruption longer than 40 ms but should not be triggered for one less than 20 ms. Also, in all cases the source is to be modulated to prevent the receiver being affected by another source, whether accidental or deliberate.

In order to stop the receiver responding to an incorrect IR signal, synchronization techniques are used to ensure that the transmitter operates only with its intended receiver, with the system set so that the beam is monitored using time-based multiplexing. This is important because single transmitters are now rarely used as they were susceptible to debris blowing into their path or birds crossing the beams. Multiple beam paths are now used, and these must be broken simultaneously. The beams can also use a variable-frequency technique to form a wall of cover and have their channels set to avoid cross-talk between units.

In order to stop an intruder stepping over the beams, tower enclosures are used to stack the sensors to a greater height. These beam towers are normally manufactured in aluminium to accommodate some four units stacked vertically.

Figure 3.8 shows an example of stacking and protection.

IR beams may be used internally and externally, although there are extra considerations when used outdoors, caused by extremes of weather. A unit

Figure 3.8 *IR beam stacking and protection*

used externally will have its operational range reduced by a value of the
order of 60 per cent compared with the same unit used internally. The operating range of an external detector in practical use would only be some
200 m but in stable internal applications it could be of the order of 600 m.

The active IR detector is ideal as a perimeter sensor in detecting intrusion at a very early stage. It also cannot be easily masked, and since it monitors the entire length of its coverage area it can produce an output if even
one of the beams is blocked, and prevent the system being set until the
beam is cleared.

Used for an external application the considerations are:

- *Bad visibility.* IR at 900 nm operates reasonably well through fog but in
  areas where this occurs on a regular basis it will be necessary to use an
  IR sensor with automatic gain control (AGC). This continually monitors
  for gradual changes in the signals caused by changes in weather conditions. It then adjusts the trigger level accordingly to maintain the proper
  sensitivity level for the current environmental conditions. In fine weather
  the AGC keeps a high trigger level to prevent interference from external
  light. In bad weather conditions this circuit automatically lowers the trigger level as the IR energy becomes blocked by rain, snow or fog. In the
  event that the signal is lost, an output is generated as 'trouble' to advise
  an operator of closedown.

- *Air turbulence.* This is a greater problem in hot climates and long-range applications, and it is necessary to reduce the operational range to compensate.
- *Low temperatures.* Anti-frost designed slots on the cover can allow the beam energy to pass even when the cover is completely frosted over. If the device has a high tolerance to signal loss, then even a small amount of IR energy will ensure stable operation. Beam systems can also feature heaters and thermostats to clear the beam windows from frost and ice. Heaters may be included in the optical head of the beam sensor itself. They can be placed within a tube to make use of the vortex effect. This design accelerates the flow of air over the heater and projects warm air to the position where the beam passes through the outer cover. When used with towers, one thermostat is needed for each tower and should be fitted at the bottom of the tower.
- *Power supplies.* These may need upgrading when heaters are used, and can be placed at a number of points in the larger system and closer to the sensors.
- *Anti-tamper.* The system must be fully tamper protected, and when towers are employed a top tamper switch should be used to prevent the tower being used as a climbing aid so that downward pressure will activate the alarm.

Installation

Infrared beam barriers are ideal for protecting walls, entrances and window apertures and are easily installed. Miniature versions can be confidently used in the domestic environment or be extended to the commercial and industrial market. They can be used with any control panel as they have alarm, tamper and often also relay contacts so that heavier loads may be controlled when the device is alarmed.

Photobeam alignment need not be a costly, time-consuming job. Using a reputable system with alignment status communication, the beam alignment level is visually displayed using LED indicators on both the transmitter and receiver (Figure 3.9). By simply aiming both units and watching the LED indicators, accurate and reliable alignment is easily achieved, and one-man installation and maintenance is possible. The alignment status at the receiver is optically transferred to the transmitter, and is indicated by LEDs. To align the transmitter one simply monitors the LED level while adjusting.

The lenses themselves are adjusted to the correct alignment by means of screws for either the vertical or horizontal direction.

Precautions. Strong light from the sun, and headlights shining directly on the transmitter/receiver, should be avoided. If strong light stays in the

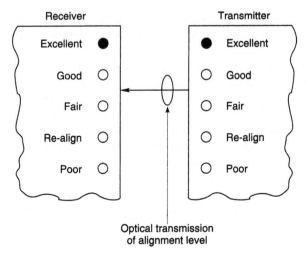

Figure 3.9   *IR beam alignment*

optical axis for a long time it will affect the life expectancy of the sensor. Do not install the unit where it may be regularly splashed by dirty water. Obstructions between the transmitter and receiver should be removed, and installations should never be on unsteady surfaces.

Microwave

The active IR beam interruption device is an extremely stable detector system that can be confidently used for internal as well as external applications. Although it can be affected by extreme weather conditions when used externally, more modern devices are very tolerant of environmental conditions, and the IR sensor therefore has widespread use. The microwave detector when used outside is not affected by elements such as fog, rain or frost but is otherwise a highly sensitive device affected by moving grass, bushes or trees, and is difficult to commission as it is not so easy to determine the actual range being covered.

In practice the microwave sensor used as a beam interruption device features the same technology as that of the movement detector described in Section 3.6, but it radiates its beams of microwave energy from a transmitter to a receiver which is in direct line of sight. The receiver compares the energy level with that transmitted, and if a significant difference in wave form, amplitude or magnitude is found for a given time window, it goes into alarm. This is in contrast to the use of the Doppler effect employed by the movement detector where internal building surfaces can be used to reflect the energy. This Doppler principle is not usually adopted for outside duty but occasionally single heads in conjunction with reflectors are found.

The properties of the microwave beams are governed by the type of transmitter specified although in general they do have a far greater depth of coverage than the IR type, but the field can drift and they are best used within a boundary or physical fence to ensure that only specific unauthorized access is detected. For this very reason, microwave detectors are often classified alongside other fence detection systems used close to and within perimeter boundary fences. These provide a secure area up to a distance of 200 m and should be installed on long, straight boundaries.

In cases where there are ground undulations and there is concern that an intruder could crawl under the beams, then beam shaping or phase sensing is applied. This process uses vertical or horizontal aerials to give increased ground cover or extended ground and high-level cover. The method of installation of these aerials varies with the manufacturer of the equipment.

The type of transmitter used will depend on the area that is to be protected. Vertical patterns have an energy form based on a parabolic trough which is very broad and gives protection against crawling. Although pole mounted they are only sited some 1 m above ground level. Short-range detectors are used to cover gates and entry points. The pattern is cylindrical. Long-range beams can extend to 300 m at the expense of beam width, but it is more usual to install several midrange sensors.

It is difficult to generalize further on microwave beam patterns because of the wide available range and the differences in mounting the transmitters and sensors.

The student is prompted to understand the role of the microwave sensor and how it can readily be used as a detector to trigger closed circuit television and lighting or as a first-stage detector. Further reference on this subject can be made to BS 4737: Part 3: Section 3.4: 1978, 'Specifications for components. Radiowave Doppler detectors' and BS 4737: Part 3: Section 3.12: 1978, 'Specifications for components. Beam interruption detectors'.

3.6 Movement detectors

These are so called because they detect movement in a protected area. The passive IR detection device (PIR) is used in high volume in the intruder alarm industry, and so it is important to understand its method of operation. Movement detectors are also sometimes called space detectors or volumetric detectors.

Passive infrared devices

All objects emit IR energy to varying degrees. The heat energy radiated by certain objects such as the sun is readily detected, but that radiated by other bodies may be so insignificant as to be almost undetectable. Absolute zero, $-273°C$, is the point at which an object has no heat energy and the electron movement in the atom stops. All objects in practice have a temperature above absolute zero, and thus possess electron movement: it is the movement of electrons inside the atom which causes IR radiation. The heat radiation pattern of a material is specific to its own chemical composition and temperature.

Heat can itself be transferred in different ways, but IR energy can only be transmitted by radiation – the transfer of energy emitted by one body through space and then absorbed by another body. IR radiation is a form of electromagnetic radiation, and this has similar properties to light: they both travel at a speed of 300 km/s in a vacuum, and generate heat when absorbed by a body, but they do possess different reflection, absorption and penetration characteristics. Like light, when IR rays strike an object they are either reflected or absorbed. It is the frequency of the electromagnetic energy and the composition of the object that determines the exact degree of reflection, absorption and penetration. The human body emits energy that peaks in the far-IR region at a wavelength of about 10 mm, and the PIR detector is engineered to effectively detect sudden changes in far-IR radiation levels and produce a corresponding electrical signal. The PIR device will compare this signal to the signal characteristics of an intruder to eliminate false alarms and improve the reliability of the detector. All PIRs consist of three fundamental components:

- optics to control the vision;
- sensor elements which generate electrical signals from IR energy;
- signal-processing circuitry to evaluate the signal and trigger a function such as tripping a relay.

PIRs are passive in the sense that they do not emit energy but are receivers of far-IR rays. Without an optical system to selectively focus the field of view, the PIR would have just one very wide and insensitive zone. Like a camera without a lens, it would only detect very close objects. Plastic Fresnel lenses or mirror segments are used to create optical windows, and it is these which control the detection area, zone size and sensitivity of the PIR.

The detection zone is created by the optical window receiving the IR rays and then focusing them onto the sensor element. In addition, the optical window also blocks out IR radiation from unwanted areas. The height, width and shape of the detection zone are determined by the sensor element, focal length and the distance from the detector where the zone is measured. The angle or rate of the expansion of the zone is determined solely by the focal length of the detector. A long focal length creates a narrow field of view for the PIR while a short focal length creates a large zone more quickly. The detection zone becomes larger at a constant rate whilst the zone shape is determined by the shape of the sensor element. The zone size should be the size of the target (Figure 3.10). Compared to smaller zones, a human-sized zone is just as sensitive to human targets but less sensitive to small disturbances such as mice and other rodents or heater vents.

The window size determines how much IR energy is collected but it has no relationship to the size or shape of the zone. Since large windows collect more IR energy, they can produce more clearly defined signals than small windows, and are thus more desirable for long-range applications. High-quality lenses or mirrors are essential if a sharp focus is to be created on the sensor element and thus produce a clearly defined detection zone. A clearly defined detection zone is important because it provides clear signals and high sensitivity.

The sensor elements are pyroelectric elements, and absorb the IR energy focused onto them, transferring it into heat. When the amount of IR energy they receive changes, the elements change temperature and create electricity. It can therefore be seen that the heating and cooling of the pyroelectric elements create signals which are processed by the detector (Figure 3.11). Although PIRs do detect sudden temperature changes in the environment, the objects that they detect usually have not changed temperature.

PIRs are less sensitive to fast-moving objects at short distances and to slow-moving objects at long distances.

Figure 3.10   *Zone size*

Figure 3.11   *Electrical signals generated by a pyroelectric element*

A person walking into a small zone would fill it so quickly that the actual duration of the temperature change would be too short to produce a significant signal. Equally, if a zone is too large an intruder may enter it so slowly that a trigger signal is not produced (Figure 3.12). It is thus clear that human-sized detection zones provide the best detection performance. Well-designed long-range detectors have extra short-range zones in order to ensure sufficient sensitivity close to the detector (Figure 3.13).

Dual opposed sensor elements

Detectors which use a single pyroelectric element generate signals in response to any change in IR levels. This, however, is sensitive to background temperature changes, draughts and other environmental disturbances. The development of the dual opposed sensor element (Figure 3.14) reduced the effect of causes of false alarms by combining two elements, each with opposing charges, within one detector. Each of these elements creates a signal (either positive or negative) when it changes temperature, but when both are affected simultaneously the positive and negative charges cancel and no signal is generated. This characteristic suggests an important installation consideration: dual-element sensors will be more

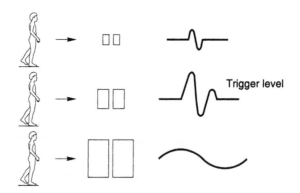

Figure 3.12   *PIR trigger level*

Figure 3.13   *Long-range detector zones*

sensitive to movement across the detection zone (alternating + and – signals) than to movement towards the detector (simultaneous + and – signals). This is most noticeable in high temperatures when the background and target temperatures are similar.

A dual opposed sensor takes each radiating zone and effectively divides it down the middle to give a positive section and a negative section. In normal use an IR source must appear in an active zone on one section and then appear on the other section within a given time scale. The sensor processor monitors each section for this energy level passing between sections, hence the detector can only respond to movement within a given time scale and will disregard a static heat source which is effectively on both sections at the same time.

The progression on this was the quad system, which uses the two fundamentals of the dual opposed element. The two dual-element outputs are fed to a signal-processing unit which will go into alarm only when two signals from the quad system exceed a pre-determined threshold level.

The object of introducing the dual opposed or quad system was to introduce resistance to false alarms caused by temperature changes within a protected area and ultimately lead to the use of a time frame pulse count. This uses a counter to record a specified number of trigger strength signals within a given amount of time before activating the alarm relay. If the required number of signals are not generated within the time frame, the counter returns to zero.

Most PIR detectors now adopt pulse counts as standard, and the number of counts required can be selected by the installer and is normally achieved by the use of a jumper on a circuit board.

Observations

- *Types:* wall mounting; ceiling mounting.
- *Range:* long-range corridor – up to 30 × 2 m (Figure 3.15a); volumetric – typically 15 m, 100° (Figure 3.15b); curtain – typically 15 m (Figure 3.15c); ceiling – typically 15 m, 360° (Figure 3.15d).
- *Activation:* radiation of IR energy reaching a pyroelectric sensor.

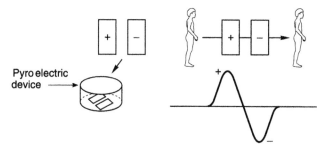

Figure 3.14 *Dual opposed sensor*

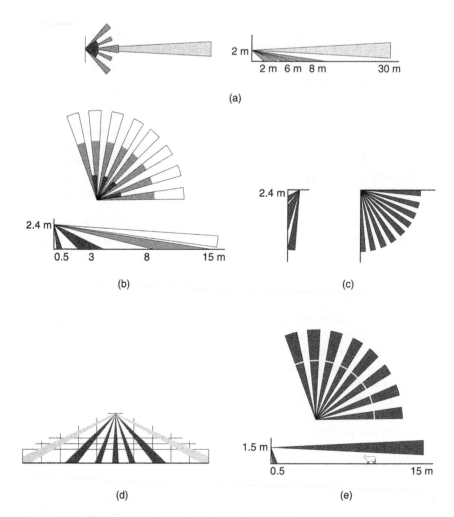

Figure 3.15   *PIR patterns (typical). See text for explanation*

- *Variant uses:* mid-risk security; protection of open areas.
- *Not recommended:* for areas with pulsating heat sources; for areas where animals may be present (pet alley lenses may be suitable (Figure 3.15e)).
- *Advantages:* efficient, inexpensive, reliable, aesthetics of high value.
- *Disadvantages:* with high pulse counts the sensitivity can be low.

Standard: BS 4737: Part 3: Section 3.7: 1978, 'Requirements for detection devices. Passive infra-red detectors'.

BS 4737 defines the area being covered as that of a person of 40–80 kg moving laterally through a distance of 2 m at any speed with the rate of rise at the sensor being greater than 0.1°C/s.

Considerations

- The vast range and availability of PIR detectors ensures that the installer can easily source a detector for most applications.
- Ceiling-mounted detectors are valuable in the protection of areas where wall space is at a premium and shelves reach ceiling height. The ceiling mounted unit is not blinded easily by the stacking of goods against the walls and up to the ceiling level. Many volumetric PIRs come with ceiling-mounted brackets, but these are not to be confused with true 360° ceiling-mounted units which project their cover directly downwards.
- PIRs are more sensitive to movement across the detector rather than away or towards it, and hence the reason for corner mounting of volumetric devices, as they more efficiently fill a space when mounted in such a way.
- Rooms containing multiple panes of glass are a hazard. The PIR zone will not penetrate glass, but visible light having passed through glass turns to heat when absorbed by a body. Detectors should never be subject to direct light, and even reflected light can cause a problem if deflected onto the PIR. Sunlight falling directly onto a PIR will certainly cause unwanted activations, and care must be exercised if the device is in an area with mirrors or highly polished metal, as they will reflect IR energy.
- The PIR must be sealed against the entry of insects or draughts by filling entry holes with silicone.
- Rodents or birds will have increasingly strong effects on the device as they become closer to it.
- Some PIR detectors are fitted with pet alley lenses to avoid pets, and lenses with creep zones to stop an intruder creeping under a detector.
- The connecting wires between the control panel and the PIR can become 'antenna influenced' and transmit radiofrequency interference. Twisted wire will reduce the problem, and rejection can be improved by placing 30 μF capacitors across the detector and control panel.

Ultrasonic

Not as popular now, the role of the ultrasonic detector has been largely superseded by the PIR. The ultrasonic detector uses the Doppler effect, the apparent shift in frequency which occurs when energy waves (whether sound, light or some other form) are emitted by a moving object. When this object is moving towards you the waves which it produces will always appear to be of a higher frequency than they actually are. If the object is moving away then there will be an apparent drop in frequency. If a room contains a source of sound, then the sound being reflected from a person moving towards the listener has an apparently higher pitch, whilst the sound being reflected from a receding target has an apparent lower pitch.

The effect of a person moving around a room is not adequate to produce a significant change in pitch that can be detected by human hearing, although this Doppler shift can be picked up by electronic devices.

The ultrasonic detector utilizes very high-frequency sound, higher than that which can be detected by the human ear – ultrasound. The frequencies of detectors vary between 23 and 40 kHz (16 kHz is the upper range of human hearing). The person being detected is thus unaware of the detector's operation although the normal motion of the target person creates a perceptible Doppler effect, the detector circuit having been engineered to detect variations in pitch of the sound when reflected from a moving person.

The detector circuit itself consists of two parts, the transmitter and receiver, which are both on the same circuit board. The signal from the transmitter to the receiver is achieved indirectly, such as by reflecting it off a wall. If a moving object causes some of the transmitted ultrasound to be reflected to the receiver, then a shift of frequency will be produced on this signal. The receiver is, however, designed to detect the presence of more than one input signal frequency before operating a relay.

A high-gain amplifier is utilized at the receiver input, and the output from this is fed to an envelope detector which is similar to that used in AM radio receivers. If there is only one input signal frequency then the only detector output will be of a direct-current bias. However, if a moving object causes a Doppler shift in part of the signal this will produce an audio output from the detector. This is formed by the main signal and the Doppler-shifted signal reacting at the detector to produce a beat note, as will any two frequencies that are fed to an envelope detector. In practice the output frequency from the envelope detector will be governed by the rate of speed at which the object is moving. It will vary from a few hundred hertz down to a few hertz. The output from the detector is amplified and then rectified and smoothed to produce a direct-current signal which subsequently operates a latching relay driver circuit. Figure 3.16 illustrates the basic arrangement.

The detection pattern of the ultrasonic detector is typically as shown in Figure 3.17. Although it is possible to also obtain sensors giving a longer and more narrow spatial projection, or ceiling mount options with 360° patterns, these are rarely encountered.

Observations

- *Types:* wall mounting; corner mounting; ceiling mounting.
- *Range:* linear – typically 8–10 m in length and 3–4 m in width; ceiling – 5–15 m at cone base.
- *Activation:* motion of a target creating a Doppler shift (effect) or variation in pitch of a given frequency.

Figure 3.16 *Ultrasonic detector*

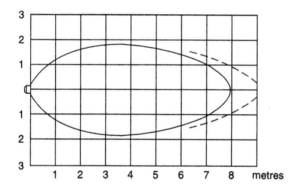

Figure 3.17 *Ultrasonic detection pattern*

- *Variant uses:* mid-risk security; stable environment needed.
- *Not recommended:* for areas with air movement or moving machinery.
- *Advantages:* can be used in large glassed areas as ultrasonic energy is attenuated by glass; unaffected by sunlight.
- *Disadvantages:* larger than a PIR with less range, and more expensive.

Standard: BS 4737: Part 3: Section 3.5: 1978, 'Specifications for components. Ultrasonic movement detectors'.

BS 4737 defines the area covered by the ultrasonic sensor using the Doppler effect as that in which an alarm is triggered by a 40–80 kg person moving through a distance of 2 m at any speed. The standard states that response should be to signals in excess of 200 ms. Also, the frequency is to be greater than 22 kHz, and the sensor must be capable of operating at 0–40°C and at humidities between 10 and 90 per cent. A warning is given in BS 4737 that nuisance or health hazard is possible from intense ultrasonic radiation, and safe limits are being considered.

Considerations

- The ultrasonic detector is more sensitive to movement in a direct line with the sensor, and is therefore generally found mounted in the centre of a wall, ideally close to ceiling height. It must be firmly fixed, and high enough to give maximum space protection, and should be angled down a little in order to reflect back.
- The ultrasonic technique is best employed in rooms with fairly 'hard' surfaces and finishes that are firm and capable of reflection.
- The ultrasonic device is directional, and so should not be mounted above a door because it is blind directly underneath (unlike a PIR with creep zones); it should be directed at a likely intruder entry point.
- The ultrasonic detector can operate in areas having large glass panels. Glass attenuates high frequencies and reduces the chance of loss or outside interference.
- Ultrasound can be generated by machinery as harmonics of the normal noise output, this may be picked up along with the sensor frequency to produce a beat interpreted as a legitimate signal. For this reason the ultrasonic detector should not be used in areas of noisy machinery.
- The ultrasonic device should also not be employed in draughty premises, as air turbulence can distort sound propagation. A similar problem can be caused by air heating, which may compound the trouble by moving curtains and drapes.
- Attention to causes of false alarms should involve the evaluation of telephones, fans, noise introduced via air ducts and fire alarm bells, although sirens are rarely troublesome.
- The use of averaging circuits, which only respond to a net change in target distance, can avoid many of the noted false alarm sources, but at the expense of sensitivity.
- Ultrasonic detectors can be affected by cross-talk between devices in the same location, and should not be placed in an area that is adjacent to another area where staff may be working, as effects such as headaches may be caused.

Microwave

Microwave detectors use the Doppler effect in the same way as the ultrasonic sensor but use radio waves that oscillate at an extremely high frequency instead of ultrasound. The microwave generator is a transducer known as a Gunn diode oscillator (GDO), which produces electromagnetic energy waves at frequencies of the order of 9.3–10.7 GHz, some 1 million times higher than ultrasound.

Microwave energy is electromagnetic and therefore travels at the speed of light. It has a very short wavelength, and because of this it is capable of penetrating many common materials, while being only moderately attenuated. Only dense materials such as metals and heavy building concrete are able to attenuate or reflect microwaves. It can be focused in the same way as visible light, and it can also be reflected and refracted, making it controllable for precise aiming at a given area.

The advantage of microwave energy is that it can be used to detect intrusion before a secure area is penetrated, but care in other applications must be employed as it will penetrate many partition walls, leading to detection beyond the area within which the detector is confined.

There are certain advantages that microwave technology has over that of ultrasound when used as a detection technique. As microwaves travel at the speed of light their energy level is so high that they pass through air with tremendous efficiency. For this reason, microwave energy will pass through air as readily as through a vacuum, and so it does not need to rely on air pressure for transmission in the same way as sound waves; thus, air currents will not disturb microwave sensors causing false alarms.

Using the same logic as the ultrasonic device the receiver and transmitter are built into a common housing, and some direct radiation from the transmitter along with that from objects in the field of view will impinge on the receiver.

Although microwave energy will pass through many materials it is found that the human body does not allow transmission because of the mass of water that it contains, which attenuates and reflects the microwave energy source. In order for the Doppler effect to be produced some of the transmitted energy must be reflected and some of it compressed by the object moving in a plane towards the detector, hence intruders must move within boundaries that will contain some of the transmitted energy and also reflect some of it.

The microwave detector is constructed to have an opening or microwave cavity (wave guide) of precise dimensions to focus the microwave energy onto the area in need of protection. An amplifier is used to process signals that result from Doppler frequencies to provide an increase in amplitude or signal strength. The amplifier must have a high gain and a low signal-to-noise ratio so as not to amplify noise.

Using signal pulse count technology will enable the microwave detector to capture an intruder moving a short distance very quickly as effectively as it will detect an intruder moving a short distance very slowly. This is achieved by means of an accumulator which stores characteristic Doppler shifts occurring during a given time period. If at the end of the given time window the sum content of the store is over a determined threshold an alarm is raised. It must, however, be remembered that the microwave sensor is an extremely sensitive device because of the high-frequency stability and short wavelength of microwaves.

Figure 3.18 shows typical protection patterns.

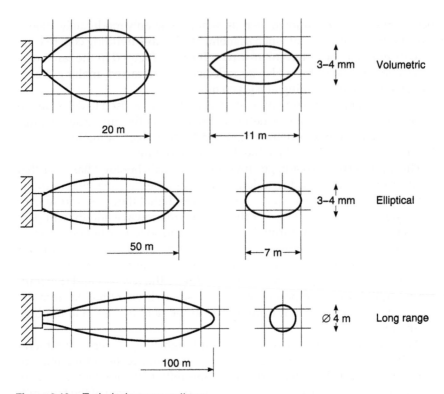

Figure 3.18   *Typical microwave patterns*

Observations

- *Type:* single transceiver.
- *Range:* linear – typically 20 m in length, 11 m in width, 4 m in height; long range – up to 100 m at a diameter of 4 m.
- *Activation:* motion of a target creating a Doppler shift of electromagnetic radiation of extremely high frequency (microwaves).
- *Variant uses:* to protect large open spaces or long corridors and passageways.
- *Not recommended:* in small areas; for operation in temporary buildings.
- *Advantages:* sensitive and controlled easily; can be used to penetrate light materials.
- *Disadvantages:* larger than PIRs and more expensive; care needs to be exercised due to its penetration capability.

Standard: BS 4737: Part 3: Section 3.4: 1978: 'Specifications for components. Radiowave Doppler detectors'.

BS 4737 defines the area covered by a microwave sensor using the Doppler effect as that in which an alarm is triggered by a person weighing 40–80 kg moving at between 0.3 and 0.6 m/s through a distance of 2 m, or 20 per cent of his or her radial separation from the sensor. The sensor must only respond to signals over 200 ms.

Considerations

- The principal use of the microwave sensor is for protection of wide, open spaces. With its long range and wide volumetric pattern it can be used to great effect in warehouses, halls and galleries. Its long range can be as great as 100 m, but its energy can also be focused into beam shapes by means of steel caps with different opening shapes. These caps are called horns or antennae.
- The microwave sensor must be securely fixed onto a surface free from vibration. It should be located away from partitions, false ceilings and heavy doors.
- Avoid mounting microwave sensors close to transformers or other resonating equipment.
- Never allow the microwave sensor to view fans or louvres. Fluorescent lights radiate electromagnetic energy, and the microwave sensor should not be directed at them. New-generation highly reflective louvres for use with fluorescent lights in areas with VDU screens are a further hazard.
- Do not aim the detector at lightweight partitions or glass as the protection will extend through these materials.
- Movement is more readily captured in a direction towards or away from the sensor element.
- As with the ultrasonic detector, the range is influenced by soft surfaces absorbing energy. Blind spots will be generated by obstructions such as cabinets, more so when made of metal. Cross-talk can also occur.
- Microwaves do not have sharply defined edges, and the definition of range is hence more difficult.
- When considering a manufacturer's stated range for a sensor, it must be noted that this can vary greatly in an area of free space with only hard surfaces and when an object of high mass is travelling at a high speed and is highly reflective.
- Microwaves are difficult to mask and cannot saturate an area in the same way as ultrasound, but they are not affected by the movement of air, or changes in humidity or temperature.

False alarms

We may summarize the major false alarm sources of the three movement detectors by reference to Table 3.2.

Although we may have looked at the principal protection patterns of PIR, ultrasonic and microwave detectors there will be found many variant

Table 3.2   *False alarm considerations*

Consideration	Sensor type		
	PIR	Ultrasonic	Microwave
Turbulence, draughts	+	×	✓
High-pitch sound	✓	×	✓
Heaters	×	+	✓
Drapes moving	✓	×	+
High humidity	+	+	✓
High temperature	×	+	✓
Reflections	✓	✓	×
Sunlight	+	✓	✓
Movement beyond the protected area	✓	✓	×
Vibration shock	✓	×	×
Water moving in pipes	✓	✓	×
Small animals	×	×	×
Mutual interference/cross-talk	✓	+	+

+, slight problem; ×, major consideration; ✓, no problem

patterns. These patterns are so designed as to avoid false alarm sources. One way of doing this is allowing the lenses to alter the field of view.

Combined detectors

All detectors are subject to some form of environmental problem. Changes in the environment around the sensor often cannot be tolerated or ignored, and this leads to unwanted activations of the device. Combined detectors adopt two different technologies within a single head to negate the problem and to confirm intrusion. There are a number of different combined detectors available, but the most popular is the PIR/microwave detector, often referred to as a dual-technology device. However, the reader will certainly also encounter combined PIR/ultrasonic, PIR/acoustic, PIR/breakglass, and PIR/air pressure change sensors.

An alarm condition can only be made in response to a stimulus to which both detection techniques will respond. The alarm signals from each detector are combined by a logic circuit in such a way that they must alarm within a preset time period before the alarm signal is given.

Combined detectors give excellent detection performance, and are ideal for use in remote sites where false alarm call-outs would be expensive and troublesome. High-risk areas, potentially troublesome environments and sites requiring coverage over large areas are well served by this type of device. New-generation devices are becoming popular because of their resistance to false alarms and because they are approaching the PIR sensor on price and size. They are therefore poised to enter the residential and light commercial building market.

The installer must consider use of the dual-technology sensor carefully, since if there is no environmental problem there is no need to introduce a more complex detection device which must increase, even if only marginally, the possibility of component failure.

PIR/microwave

The dual-technology detector in this case relies on PIR detection to sense body heat and the microwave Doppler effect to sense movement. Only when both types of sensor are triggered simultaneously will the detector go into alarm. This significantly reduces the possibility of false alarms as both technologies are fundamentally different in their detection technique. The very early microwave Doppler units were made using cast horns, and for this reason were quite bulky. The lowest frequency they could be made to work at was of the order of 10 GHz, known as the X band. Recent advances in technology have allowed the use of conventional discrete electronic components for high-frequency applications, permitting operation in a new S band channel at a frequency of 2.445–2.455 GHz.

The use of precision microstrip technology has produced a highly efficient microwave source enabling operation at very low power levels, typically 20 mA at 12 V DC. The stripline technology is incorporated into a wire antenna, providing advanced signal-processing techniques in a low-cost stripline.

Custom-designed solitary ASICs (application-specific integrated circuits) using both analogue and digital circuitry have been adopted. The analogue circuitry handles the signals from the pyroelectric or the microwave sensor, while the digital section provides the logic functions, such as pulse count, signal discrimination and timing. Connections between the components in the ASIC are much smaller than in earlier types of sensor, making the product less susceptible to radiofrequency interference and electrical supply transients.

The protection pattern will be seen to vary enormously between manufacturers, but in practice will certainly cover volumetric, long-range and ceiling mount variants. The microwave element may also have a two-position range adjustment selected by a jumper to give full or half projection. Separate LEDs can signify actuation of the separate elements, and a control line to the panel can inhibit the LEDs if necessary to disguise the coverage from others.

These sensors are only slightly larger than a conventional PIR detector and can therefore enter the domestic market where the single-technology microwave element is unacceptable. The range is, of course, governed by the combined elements, and is therefore reduced from that of the microwave-only detector.

PIR/ultrasonic

This type of device can be found, but has been largely superseded by the PIR/microwave detector.

PIR/acoustic and PIR/glassbreak

In these devices the digital logic circuits in the detector monitor both elements and only generate an alarm if acoustic or glassbreak detection is followed by PIR detection. These devices are thus used where there is a need for the former condition to be verified as entry (by the PIR element). These devices can satisfy the requirements of perimeter and space protection by single-unit installation.

PIR/vibration

This type of sensor confirms intrusion following a designated level of shock or impact.

It must be understood that no dual-technology sensor is classed as sequentially confirmed, and reference to event verification (see Section 6.9) must be made to understand the difference in detection criteria under the ACPO policy.

PIR/air pressure change

This is a new generation type detector using pressure sensing technology to generate first alarm when a forced entry occurs, followed by PIR technology to generate a confirmation signal when movement is detected within the covered zone. It combines PIR and pressure wave detection techniques in one housing, and with completely independent outputs, so is classified as a sequential confirmation detector. It is therefore aimed at helping installers comply with ACPO 2000. The pressure wave technology can differentiate between a forced entry and normal pressure fluctuations being immune to thunderstorms and air conditioning. It may be installed in fully enclosed rooms that are effectively areas of up to some 700 cubic metres ($15 \times 15 \times 3$ m). The detector has different pressure sensitivity settings but maximum efficiency is achieved in rooms with no vents, flues or poorly fitting doors or windows. The pressure sensing technology is engineered to respond to forced entry violations of doors and windows. The PIR element is more traditional using a dual element pyroelectric sensor with signal processing. Separate LEDs are used for walk testing the different detection elements.

Pet tolerant detection

The original solution for customers who had pets but wanted an alarm system was to fit pet alley lenses to PIR detectors to effectively mask a portion of the protected area. This has already been referred to at earlier stages in this chapter. Such a practice provided a certain level of stability on the assumption that the pet stayed very much at ground level as this was classed as the dead area. However, progression in pet tolerant or pet immune sensors has led to the widespread use of less basic solutions and the introduction of purpose-designed detection devices. However, since these detectors are available in many different guises they must be installed in an individual way and exactly to the manufacturers' requirements with correct ranges and other criteria satisfied or they will fail to satisfy their purpose. In operation these sensors have the ability to disregard the presence of an animal up to a given weight or size. This is achieved by using either PIR technology with the intelligence to discriminate between genuine intruder activity and superfluous animal activity or by combining it with microwave detection so that it operates as a dual detection device with enhanced stability.

Detection devices from different manufacturers will claim varying levels of pet tolerance and varied criteria for performance, therefore it is necessary to consider each installation in its own right and select a sensor appropriate to the application. Size of animals, environment and specific site needs all have an influence on the product selection. The considerations that the installer must be alert to are typically:

- Range/coverage: this may be selectable or preset.
- Settings for alarm sequences and sensitivity.
- Time windows for pulse counts or for microwave activation following PIR detection.
- Degree of pet immunity required. Dogs, cats, random flying birds, rodents or caged birds. Normally dogs of up to and within the range 40–100 lb can be accommodated.
- Distance that animals must remain from sensor element. Generally in the order of 1.8 m.
- Stated sensor mounting height within the protected premises. Specific to the product and vital if pet immunity is to be achieved.
- Height at which animals should not be able to get above.
- Need for an active creep zone without compromising pet detection. Certain look-down zones may not be used in pet tolerant applications.
- Requirement for the unit not to view a staircase which animals can climb.
- Appreciation that there may be a lack of sensitivity of device and catch performance at extremes of range.
- Appreciation of the stated ambient temperature range of the detection device and the influence on the tolerance of pets at the different ambient levels.

- Tolerance of small, slow-moving animals in relation to larger animals.
- Selectable sensitivities for single or multiple pets.

3.7   Inertia and vibration detectors

The object of this section is to promote an understanding of those devices capable of recognizing seismic vibrations caused by attempted illegal entry into protected premises. Forced entry will always cause vibration of the surrounding structures, and the detection devices we are interested in are all capable of being fitted to window/door frames or any type of surface or panel that is in need of protection.

We must realize that in the security industry there are a number of terms applied to detectors fulfilling the same purpose. Those most encountered are 'impact', 'shock', 'inertia', 'vibration' and 'seismic'. Indeed, specifiers may even use different terms for similar products from different sources. Generalizing, inertia and vibration are the most used terms.

Impact, shock or inertia tends to refer to action by electromechanical means, in that a switch is activated as a result of gravity. Such a device will therefore have moving parts and contacts.

The vibration or seismic device differs in that it will use a piezoelectric element, which generates electricity from molecular vibration caused by pressure as a result of the piezoelectric crystal being affected by mechanical vibration of the surface upon which the sensor has been installed. This electrical signal is amplified and triggers the alarm state.

Inertia

The original inertia-type sensor comprised a gold-plated ball seated on a pair of contacts to form a normally closed circuit. An impact would displace the ball, causing the circuit to go open. This type of detector was interfaced with an electronic analyser with processing and selection circuits which determined the magnitude of the response of the ball to impact to be inputted to the control panel. In multisensor applications all the sensors were controlled by the same analyser.

With new-generation control panels a feature often found is a gross attack function. This gross attack level refers to the sensitivity of a zone to a sensor activation. The higher the gross attack level the less sensitive a zone is to an activation. The response of a zone to a single impact is hence dictated by its gross attack level. The sensitivity of the inertia sensor can therefore be set from the control panel if it has such a gross attack function.

Mass inertia technology is suitable for use on any surface. Its principle of operation is that an electrical circuit is kept normally closed by a weight, but under shock conditions the main body of the shock sensor is caused to move whilst the weight remains stationary owing to its inertia. This causes

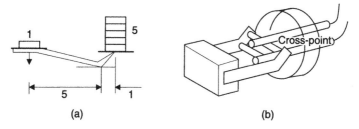

Figure 3.19 *(a) Mass inertia; (b) Cross-point*

electrical contacts to be momentarily opened. These contacts may be gold-plated spheres, washers or discs which directly rest upon two or more gold-plated contact points.

The construction of the device should be such that in order to prevent false alarms from low-frequency movement, caused by wind or traffic, the weight is as light as possible so that it moves with these vibrations. However, in order to give high contact pressure this weight must also be as heavy as possible. These are obviously opposite requirements. The solution is to adopt a crowbar sensor technique which uses a leverage action, enabling a relatively light mass to produce at least 500 per cent more contact pressure than could be attained with a direct-acting system of the same mass yet still ensuring maximum stability and performance (Figure 3.19).

In any mass inertia sensor the actual point of contact needs to be as well-defined and as small as possible. Small-diameter highly polished contact rods which cross each other at right angles (giving the smallest point of contact) and high pressure are used.

The requirements of light mass, maximum contact pressure and high-integrity contacts all give the sensor maximum stability, and thus many of the normal false alarm causes such as wind, thunder, traffic and building vibrations fall below the detection threshold although forced entry is readily detected. Combined with the microcontroller-based signalling systems these detectors do not need adjustment to be performed by variable resistors; however, applying test signals, LEDs will indicate the correct settings. Pulse count systems will monitor variably sized small shocks and determine if they are caused by an actual entry attempt or some spurious signal. The larger the initial shock signal the longer the time window (Figure 3.20).

The device will continually monitor for shock signals, and any in excess of 1.5 ms will be stored in the memory. Further signals received within the time window are added to this and stored. If the alarm level is reached within the designated time window an alarm is generated. If the alarm level is not reached within the time window then the pulse count memory will be cleared and no alarm will result. This window can self-adjust depending on the size of the initial shock. If the initial shock is small, then so is the time window, and vice versa.

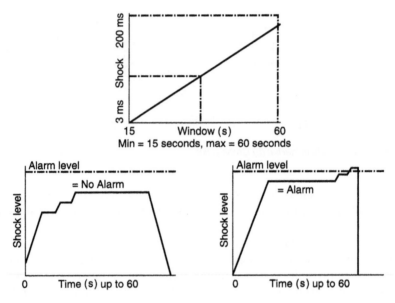

Figure 3.20   *Alarm threshold*

We therefore class this as a stand-alone device in that it has an inbuilt analyser which enables individual adjustment of all units by the integral LEDs. The sensor can then be wired direct to a control panel.

The overall size of the vibration detector is of the order of 92 × 25 × 21 mm, and it is mounted by screws applied through the back-plate. The material of the sensor casing is generally high-impact polystyrene with a white finish for aesthetic purposes. The current consumption is only some 17 mA.

Vibration

As previously stated, the piezoelectric sensor relies on the creation of molecular vibration, which in turn generates an electric potential. It uses a chip of barium titanite which contains a high volume of individual piezoelectric crystals which respond to pressure. A custom integrated circuit is used as the built-in analyser, so the piezoelectric vibration sensor is also a stand-alone device.

In practice it will be found to be of the same size as the inertia detector and to have a pulse count facility, so that when a pulse is detected a time window of a given duration is opened. If the remaining pulses (or a gross attack) are received by the unit within this period then the sensor will signal an alarm. Different LEDs can show the different pulses and alarms threshold. The pulse count can be selectable in order that the installer can determine the level and duration that the unit is to respond to. It is selected by setting a DIL switch.

In the event that a large area is to be covered and the control panel does not have the facility to show 'first to alarm' and 'subsequent alarm' when a number of detectors are to be used, the devices may be set to latch with 'first to alarm' and 'subsequent alarm' or 'any to alarm'. This requires a control signal (sw+) from the control panel to reset the alarm LED. The method of wiring varies, but with double pole it means that seven cores are needed (power, alarm, anti-tamper, plus one core for latch).

With some vibration detectors it is possible to connect a resistor in series with the latch terminal to enable 'first to latch' and 'subsequent to latch'. Reference may also be made to Figures 5.1 and 5.2.

The pertaining standard is BS 4737: Part 3: Section 3.10: 1978, 'Requirements for detection devices. Vibration detectors'.

The inertia or vibration sensor is very much a perimeter detection device.

Considerations

- Be aware of the form of attack that is most likely to be encountered on the area in question. All sensors have a good response to drilling, cutting and hammering. Specific sensors can have a good separate, instantaneous response to explosive attack.
- The typical size originally mentioned is to protect window frames, doors and panels, but larger, heavy duty variants in robust enclosures can be used as safe limpets and on strong-room doors.
- Sensors can be mounted on a variety of surfaces and fixed to glass with adhesive, but in all cases must be rigidly fixed. They should not be mounted on any structure that can vibrate legitimately or fixed to block work without first establishing that it is capable of transmitting the vibrations through it. Never fit this device to uneven surfaces or unseasoned timber. Windows that are subject to direct sunlight can flex and create vibrations within the glass panes.
- Continuous and low-frequency vibrations can readily be transmitted through any rigid structure from a remote source.
- When selected and installed correctly these sensors are highly efficient, but it is also wise to incorporate volumetric detection support when they are employed to protect external structures and openings.

3.8 Capacitive detectors

BS 4737 refers to two types of capacitive detector, namely the capacitive proximity detector and the volumetric capacitive detector.

The former device is governed by BS 4737: Part 3: Section 3.13: 1978, 'Requirements for detection devices. Capacitive proximity detectors'. It is designed to generate an alarm condition in response to the change in capacitance resulting from the proximity of a person to a protected object.

The working principle of the capacitive proximity detector is the change in electrical capacitance between a protected object and earth. Normally this is produced by the hand or body capacitance of an intruder, bearing in mind that the human body possesses both electrical resistance and capacitance.

These detectors are used to protect metal safes, cabinets containing valuables, or other metal objects. Items that are not of metallic construction can be protected by fixing metal foil strips to them.

Capacitive proximity detectors are sensitive to either an increase or decrease in capacitance, so any removal of the foil or metal object can be sensed. Slow changes in capacitance are compensated for, as is electrical interference. The unit is set up using LEDs with the sensitivity adjusted to suit the pertaining environment. Tampering with the sense lead will also generate an alarm.

Although the capacitive proximity detector is designed to generate an alarm in response to the closeness of a person to the sensor, for test purposes it is based on actual touch, but the installer should identify the area of coverage in the equipment record. These sensors are often used as the detector element in safe limpets which are attached by a keep, and they may also have a separate processing unit. If a separate processing unit or analyser is used then interconnections must be electrically protected. As with any housed detection device the unit should be tamper protected and an alarm generated if the voltage is reduced to a level that impairs the performance of the sensor.

The student should not assume that all safe limpets use this type of detection element as some versions use inertia/vibration techniques.

The second type of device is governed by BS 4737: Part 3: Section 3.8: 1978, 'Requirements for detection devices. Volumetric capacitive detectors'. These detectors are different in that they are designed to generate an alarm in response to a change in capacitance when a person enters and moves in the area or volume of coverage. The area of coverage is defined by BS 4737 as that within which the entry and movement of a person weighing 40–80 kg and walking at a speed of 0.3–0.6 m/s through 2 m causes an alarm. The area should be identified in the equipment record. The device must have a method of indication to enable the subscriber to check the cover before the system is closed.

In so far as power supplies, electrical connections and processing units are concerned, they are subject to the same requirements as the proximity version.

Although classed as a volumetric sensor, the target range of any capacitive device is restricted and only effective for, at the most, 1 m.

The volumetric capacitive sensor uses the principle of electrical resonance in that current is generated between a capacitor and inductor network which alternately stores and discharges energy. This sets up an oscillating current or tuned circuit, and if this is then influenced by a change in inductance or capacitance, it leads to a change of resonance and a variation in the current drawn. This change in current is caused by the capacitance of the human body becoming apparent in the electrostatic field or by touching the tuned circuit.

3.9 Acoustic detectors

The acoustic detector is essentially a high-security device found in bank vaults and strong-rooms or heavily structured areas which are quiet when the alarm is set and are not in close proximity to any noise sources.

The acoustic detector is activated by audible sound energy as a result of intrusion or early attack. For this reason they are better served if installed in an area with hard reflective surfaces rather than in a room with soft furnishings that will absorb the sound energy.

They are generally ceiling mounted and positioned at the centre of the volume being protected.

The acoustic sensor is covered by BS 4737: Part 3: Section 3.6: 1978, 'Requirements for detection devices. Acoustic detectors'. According to BS 4737, these devices should have a stated or adjustable nominal ambient sound level and only signal an alarm when the characteristic frequencies give a sound level greater than 15 dB(A) above this nominal value. They must also have a maximum trigger level under normal circumstances that is not greater than 85 dB(A) but must generate an alarm for any sound in excess of 120 dB(A) which lasts longer than 100 ms. They must also signal an alarm if any trigger level sound is present for more than 5 seconds in any count period of 30 seconds.

The principle of activation is based on the recognition of noise, or the vibration of the surrounding air, by means of a stable microphone, filter and integrator which process the information to discriminate between likely intrusion and sounds in a different frequency band.

Noises that are generated in a mode signifying attack of movement in a carefully controlled space are of high frequency and of the order of 1500–6000 Hz. Low-frequency sounds which occur as a result of vibrations outside of the protected area but can appear as airborne noise within this area are ignored, as are the higher-frequency whistling-type sounds. Popular models are single source which are connected direct to the control panel, but models that may have outstations can also be sourced, allowing a number of sensors to be connected via an analyser.

In order to filter out false alarm signals, dual sensors can compare noises in the protected area with those noises outside of but adjacent to the area covered. Models with discrimination of different frequencies can satisfy any particular ambient need. Pulse count when employed will sample alarm sounds within a given time window before generating an alarm. It may then be possible, as an example, to recognize a high-amplitude single pulse representative of an explosion followed by sounds more characteristic of persistent hammering to finally gain access. By means of amplification the sensor can also detect frequencies generated as a result of a known form of attack, and this may then be amplified. This can refer to scraping-type noises which in relative terms are not loud but may occur during removal of cladding materials or even building mortar.

The clear advantage of the acoustic sensor relates to its capacity to signal an alarm before access to an area can be gained. It was often used as the sensor in the safe limpet before the present-day use of vibration or capacitive technology.

3.10   Futuristic detection features

Everyone who is involved with intruder alarms will find that the technologies are changing to suit innovative and more effective products in line with the European requirements. In addition to this, as our housing and industrial and commercial buildings change in construction, the role of intruder detection must change also. Manufacturers will continue to develop new products and it becomes difficult to list all of the new additions which offer an array of new features. The future promises detection devices that are of low power consumption, compatible with all conventional security systems and iD technologies and have excellent false alarm immunity.

Protective switches will remain inherently reliable and continue to have a role to play in perimeter protection and for the monitoring of doors and other openings. EN50131-1: 2004 (section 8.7.1) requires all terminals to be located within the component housings. New-generation magnetic contacts feature terminals enclosed within a tampered housing with pre-installed end-of-line resistors for connection to fully supervised loop and double-pole systems. These tampered enclosures can use spring action microswitches again with pre-installed end-of-line resistors.

Digital shock detectors may include a magnetic contact adding an additional security feature to the device when installed on doors or windows. The shock sensor of itself may use a tri-coloured LED for simple and reliable adjustment. The LED can indicate over-sensitive, under-sensitive and correct adjustment to address the challenge of consistent and reliable shock detector calibration.

However, movement detectors will continue to make up the nucleus of the intruder alarm industry and these will increasingly include adjustable anti-masking detection to stop a mask being placed on the sensor window. This anti-masking will recognize paper, sprays, lacquers, clear sticky tape and card or cardboard boxes. Anti blocking which is a high-security mode will also become more of a feature as this provides detection in the day mode of objects placed close to the detector to obstruct their field of view.

Such blocking will be recognized by the control equipment and the offending objects can be removed before the system is set.

Unique products with separate PIR and dual technology all included within one distinct housing will also become more widespread.

Through the use of 'interdigitated' fields of view with an ability to mask either the PIR or microwave and with selectable EOL resistors and numerous LEDs to show the detector status, these devices will help to revolutionize detection techniques. Allied to this the two different detection technologies will be non-overlapping to ensure that they comply with the sequential alarm requirements of DD 243: 2004. Therefore we will see products for the higher grades advertised as:

A unique product with two non-overlapping detectors; a PIR and a dual technology in one robust housing. Ideal for use in new systems or for retro fits upgrading to PD6662/EN50131-1 Grade 3 and compliant to DD 243.

Reference to this subject can be made to section 6.9 covering alarm confirmation technology.

Manufacturers will develop lines of new-generation, technologically advanced, aesthetically designed intrusion detectors. For movement detectors these will be standard lines, pet friendly variants with bespoke technology that distinguishes signals caused by humans from those created by family pets, and combined technologies with proprietary signal processing algorithms to differentiate disturbances in the detector field. Advances in the digital arena and the integration of microprocessor technology have ensured that the physical size of detection devices has come down. Digital technology has also allowed manufacturers to fine tune their products in order to provide solutions with greater signal processing capabilities. The result of this is greater reliability. Anti-masking and anti-cloak technology will become increasingly employed for devices to EN50131-1 Grade 3, PD6662 Grade 3, TS50131-2-4 Environmental Class II.

Profiles of movement detector alarm signals will increasingly be stored to help eradicate false alarms as these sensors will have intelligence for greater system analysis. Push connectors will appear on some products to firmly hold the wires in place to minimize installation times. We will see selectable end-of-line resistors that allow simple selection of end-of-line resistor values for alarm, tamper and fault/mask outputs. With some control equipment panels all these signals can be signalled on one zone. Foam gasket sealed optics in PIRs will form a sealed area between the sensing element and lens, stopping air flow and insect infestations in the optics reducing the risk of false alarm activations. Simply by moving a switch on the PCB, detectors will be able to operate as sequential confirmation products or as triple coverage detectors so that an arm will not occur unless all three detector fields are activated in a particular sequence.

Of necessity all intruder detection products will develop to complement the techniques that have been engineered into mainstream movement detectors. Security professionals when looking at every technology will need to consider all of the strengths and weaknesses with regard to specific criteria. This school of though must be followed through from switch contacts used as protective switches to combined technology products.

All energized intruder detection products will benefit from given factors when evaluating signals from the sensor such as:

- Shape – type of waveform the signal has
- Amplitude – how large is the signal from the sensor
- Energy – how powerful is the signal source
- Time – how long are the signals present
- Frequency/multiple events – signals constant or intermittent

Future intelligence will be incorporated in most devices for system analysis with immunity thresholds set above the levels of electromagnetic interference. All products will have installation test modes and memory storage. As an example a breakglass detector will have stages to prove it can respond to the high-frequency glass shatter and then recall the low-frequency flex of the glass.

Nevertheless before committing to high volume changeover of products the engineer must consider a number of factors to ensure that new futuristic products have been fully developed and their reliability is proved. If this was not followed through the end result would be unpredictable.

The engineer must do certain things when moving to any new futuristic product that they are not familiar with, and always take the following into account:

The probability of detection
The vulnerability to defeat
The false alarm rate
The nuisance alarm rate (external detection devices)

Having considered these four conditions notes can be taken and an appraisal made. This is to be related to such selection criteria as follows before introducing the product into full installation practices:

Selection criteria	Notes
Grade	Risk grade established
Environmental class	Use in area
Protection type	Protective switch/PIR etc.
	Is range restricted or changed by the new features introduced?

Detection	Is the capture ability compromised by the new false alarm prevention features incorporated?
False alarm immunity/ Product history	Current devices used have good false alarm history. Can new devices offer a significant improvement
Manufacturers returns policy for new product	Similar to original detection devices or are different conditions attached?
Interchange ability with existing installed units	Consider easy retrofit and use with conventional and iD systems.
	Can the new proposed product take over from the original in total or do spares of original devices still need to be held on stock?
Additional features	Will additional features cause any confusion with time honoured installation practices?
Budget	Cost restraints balanced against probability of superior performance.

3.11   Perimeter intruder detection systems (PIDS)

To this point we have considered beam interruption detectors such as active infrared and microwave. These are often used in perimeter detection systems for external use. A number of other technologies are also used for outdoor perimeter intrusion detection but these tend to be rather more specialist. In practical terms these are often used to secure large areas and then to start up CCTV, lighting or to give pre-warnings. They may also be connected to voice warning systems to notify intruders that such persons are in the process of being monitored.

It is a challenging field to select outdoor perimeter detection because the areas to be protected are not so easily defined as those for internal protection. In addition the detection devices themselves must cope with animals, changing ambient levels and often difficult weather conditions, particularly snow and ice, and yet not cause spurious operation. For this reason and because of the specialist nature and the range of customized options we can only consider some of the other PIDS methods that are used and in more general terms.

All of these systems need to be assessed taking into account the nuisance alarm rate as a priority, otherwise the system will have little credibility:

As we know active infrared and microwave are based on beam interruption.

The other techniques we need to appreciate are as follows.

Electric field sensors operate by detecting a change in capacitance among a set of parallel sensor wires that are insulated. These may be attached to standoffs or installed on posts. Multiple wires can be used to create high detection zones. They have a high deterrent value. They have a medium rate of vulnerability to defeat. Good grounding of the sensor is needed to avoid false alarms.

Taut wire sensors combine barrier technology with sensors. They consist of rows of parallel tensioned wires that are capable of detecting displacements. A number of different forms of sensors are used. These may be switches, piezoelectric or strain measuring gauges. They all sense attempts to cut the wires or to climb them or even separate them. These have a low threshold of nuisance alarm rates.

Electrified barriers employ parallel wires that apply an electric shock to intruders. They may be free standing or fence mounted. Electrified perimeter systems can be used on a wide range of sites ranging from small business premises to power stations. Such systems are designed to detect intrusion and tampering attempts and deter attackers by delivering a painful but safe electric shock. Modular versions can be fitted on existing perimeter structures or independently. It is possible to link them with integrated management systems to give control over all security functions including intruder, access control, and CCTV alongside building management functions.

Surface wave These sensors use an electromagnetic field around sets of parallel wires that are supported on insulated poles. They sense human size intruders within the electromagnetic field of cover.

Seismic sensors are buried in such a way that they are capable of detecting vibration to the soil above them as caused by intrusion. The sensor consists of coils and magnets (geophones). Seismic disturbance causes an electrical current to be generated by the coils and magnets which is analysed to cause an alarm condition. They can suffer from soil movement as a result of ice and tree roots etc.

Pressure sensors are buried and detect pressure waves caused by compacting. They use sensors that are tubes filled with a pressurized liquid or they may use a fibre optic cable that responds to deformation.

Magnetic sensors consist of buried wire loops or coils. They are in different formats to detect humans or vehicles.

Surveillance radar is designed for automatic perimeter security of large open spaces and detects moving or stationary objects anywhere within a range of 10–800 metres. It is capable of working in harsh environments and lighting levels and can have built-in self test, condition monitoring and calibration. Certain products scan 360 degrees in that the radar fills a volume of space in which detection data is processed from the sensor out to the instrumental range. Intelligent assessment of the intruder threat is provided by the system software with multiple units capable of being networked together so that complete coverage can be made of expansive sites.

Video motion detection (VMD) is a CCTV technology that looks for contrast changes in a camera's field of view. They can be used to start up intruder equipment in conjunction with the sending of video images.

Originally VMD was perceived as an alternative to PIRs and other fence vibration systems and detection technologies. However, it is now taking detection to new levels. Intelligent algorithms have made VMD powerful and very capable of filtering out such things as car headlights, tree shadows, rain and extremes of snow. Target sizing has also become more accurate so that animals can be discriminated and ignored. In addition vibration that affects the camera and any shake due to its location can be rejected. Technological advances have also aided installation practices and quick set-up is now the norm. Connectivity has led to the use of other devices so it can easily be integrated with intruder. Ethernet connections can be made to help use with extra technologies via contact outputs and interfaces.

A big advantage of video motion detection is that it may be used to deliver a BS 8418 compliant installation which requires detection to actually be in the view of the camera. It can be said that as VMD only detects from the camera view then compliance is automatic by definition.

In terms of siting of the equipment the area between two perimeter fences is ideal even though it still lends itself to volumetric coverage. VMD

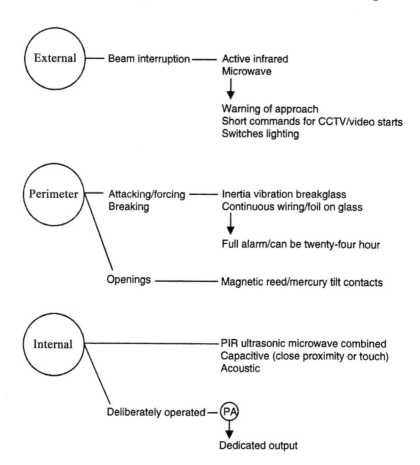

Figure 3.21 *Principal detector types and roles*

may also be used to detect the direction of a target so that exiting can be ignored if the requirement is only for approach detection of the target area.

Combined technologies using more than one technique can be used to reduce the probability of false alarms in PIDS although this may also have an effect on capture capability. In practice both technologies must alarm within a given time window. It is fair to say that for outdoor detection whether it may use beam interruption, fence detection or ground disturbance needs thorough assessment. It is becoming increasingly used with remote signalling to include CCTV to verify alarms as true.

As a conclusion to this chapter an overall indication of the major roles of the principal detectors, from external duty through to perimeter and internal applications, is given in Figure 3.21. This can only be in general terms since there is an overlap, e.g. inertia/vibration detectors although classed as perimeter devices can also be found on internal doors and openings, as can magnetic reed contacts.

4 Power supplies

At this point we need to appreciate the role and requirements for the power supply unit (PSU).

Technicians are required to install and maintain power supplies integrated within the control panel and other stand-alone power supplies that may be remotely located. There is a need to understand the regulations and principles of operation of all types.

The power supply of an intruder alarm system, in most instances, is mains electricity with a battery standby supply. In practice they may be combined or separately housed at a remote point. If the system involves remote signalling they may also be British Telecom (BT) approved.

With an increase in the integration of security systems and the linking of them with safety and management networks a greater use on the emphasis of the power supply unit (PSU) has resulted. In addition, advances in signalling and the sharing of outputs between electrical systems has further increased the need for the integrity of power supplies and the need for the installer to understand the role of these devices.

It is important to be able to relate to the requirements that exist for every situation. Equally important is being able to reference the ampere hour capacity of the standby support batteries to typical practical applications, taking account of discharge characteristics. There is also a need to establish the current rating of power supplies and when referencing remotely connected power supply units with control units the effect on control circuits must be known.

The intruder alarm engineer will also find that the introduction of the European Standards brought about more stringent requirements and a need for greater standby capacities than what had applied with the original British Standards. The requirements are now contained in EN50131-6 with recommendations held in BSIA Form No. 180.

All manufacturers of power supplies must be specific in regard to their products as it is not acceptable to rate a power supply as the full capacity of the supply. They must be rated in relation to the current available for peripherals with adequate reserve for operation on battery standby and the current taken by all peripheral products.

EN50131 also specifies four levels of access which places limitations on the accessibility to fault indications depending on the user level. This had an effect on stand-alone power supplies which are separately housed for remote locations, as it is now unacceptable for them to display external indicators of status, and this ultimately led to the use of internal diagnostic LED indication only. It is also to be noted that in all intruder alarms an indication is required when the voltage that is available from the alternative power source falls to such a level that is required for satisfactory operation of the system.

4.1   Primary supply

Although a power supply can mean a transformer, a battery or a rectifier filter with or without a charging circuit that converts alternating current (AC) to direct current (DC), alarm engineers usually apply the term to the components as a group. Most standby power supplies use rechargeable batteries as a secondary supply.

A power supply starts at its step-down transformer, which converts its 230 V AC supply to the voltage of 12–18 V AC used by most intruder alarm systems. The transformer is a device employing electromagnetic induction to transfer electrical energy from one circuit to another, that is, without direct connection between them. In its simplest form, a transformer consists of separate primary and secondary windings on a common core of a ferromagnetic material such as iron. When AC flows through the primary the resulting magnetic flux in the core induces an alternating voltage across the secondary; the induced voltage causing a current to flow in an external circuit. In the case of the step-down transformer the secondary side will have a lesser number of windings. From this transformer, power is provided through a two-conductor cable to a rectifier and filter circuit where AC is converted to an optimum 13.67 V DC. A charging circuit will be contained within the power supply so that the standby battery may be constantly charged so long as AC is present.

The power supply must always be voltage regulated and be able to hold a fixed voltage output over a range of loads and charging currents. Microprocessor components, especially integrated circuits, are designed to operate at specific voltages and are not particularly tolerant of fluctuations. Low voltages cause components to attempt to draw excess power, further lowering their tolerance, whilst higher voltages can destroy them. For these reasons the voltage should be measured at source and once again at the input terminals on the equipment point.

The critical factor in selecting a power supply is in determining the load it must support. The first step is to establish how much power will be required by all power-consuming devices connected to the supply. The length of time that the standby supply must be able to satisfy the system if the primary supply is lost is then calculated.

The primary supply is the electricity supply to the building, and which will support the system for most of the time. The secondary supply is the support system in the event that the primary supply fails, i.e. the batteries. The systems in which we are interested will tend to be powered by a transformer/rectified mains supply plus rechargeable secondary cells via a power supply unit or uninterruptible power supply (UPS). Other power supply systems may comprise a transformer/rectified mains supply plus non-rechargeable (primary) cells, or primary cells alone, but these two types are less widely used. It follows that the intruder alarm relies heavily on the mains supply, which must be a source that:

- will not be readily disconnected;
- is not isolated at any time;
- is from an unswitched fused spur within 1 metre of the PSU;
- is free from voltage spikes or current surges;
- is supplied direct to the control panel and not via a switch or plug and socket or remote spur that can fail or be switched off.

The transformer must be sited in an enclosed position and be ventilated, and must not be placed on a flammable surface. Transformers are found within the control panel itself, or in the end station in the event that the system employs independent remote keypads. Within the same confines will be found the rectifier and charger unit. The system will have either a battery charger unit (BCU) or a UPS.

The UPS has a greater ability to negate interference and surges on the mains supply, and it tends to be widely used in computer power supplies that have back-up systems. The essential requirements of a battery charger are that:

- it can recharge all batteries to full charge within 24 hours whilst maintaining the system load;
- it is internally fused, both primary and secondary;
- it is free floating and includes audible and visible indications of failure;
- it includes a voltage trigger to activate remote signalling of failure;
- tamper protection of the cover is provided;
- it has short-circuit protection with a grounded negative on the secondary DC.

As previously stated, a UPS has greater protection to interference with increased recording and monitoring. It must also feature a safety isolating transformer and have the specified output plus recharge requirements under any combination of rated supply voltage and supply frequency at temperatures between −10 and 40 °C.

The UPS will additionally have a low heat output fully rectified transformer, solid state voltage regulator, linear current regulator and high-temperature cut-out with continuous monitoring of the low-voltage alarm circuit. Mains suppression filters are used to remove transient high-voltage spikes. The intruder industry requires the following UPS units:

- that they be of sufficient capacity and recharge rate to cope with any prolonged mains isolation of the main supply related to work being done for fire safety, normal isolation or normal work on the electrical services;

- that they are located where maintenance can be easily performed;
- that sufficient ventilation is afforded to stop gas build-up on the vented battery occurring and causing damage or injury;
- that they not be exposed to corrosive conditions and that the cells be fully restrained to stop them falling or spilling;
- that the units must be marked with the date of installation.

Before considering the types of secondary supply in use within the intruder alarm area, the student may wish to pay some attention to the inspection of the mains supply and the tests that must be performed to prove it acceptable. These tests range from visual checks for cable damage to electrical proving requirements, and are covered in Chapter 8.

4.2   Secondary supplies

As a secondary supply, batteries are an integral part of every electronic security system.

Batteries, particularly rechargeable ones, are not simple devices. Although they can withstand much abuse, they will serve their purpose better if they are understood and correctly maintained.

History

Following the work by Volta in the 1790s using copper and zinc with salt water to produce electricity, the lead–acid system was discovered by Faure in 1881, and since then has been progressively refined and commercialized throughout the world. In the last 20 years, development has been more rapid in the search for improved power-to-weight ratios and new applications. Demands for portable power to supply a wide range of modern equipment have resulted in the latest addition to the range of products based on the lead–acid system – the sealed lead–acid (SLA) battery. It is this unit that dominates the intruder alarm industry.

Batteries can be classified in two broad categories:

- primary – non-rechargeable;
- secondary – rechargeable, e.g. wet cells, nickel–cadmium, SLA.

The term 'sealed lead–acid battery' is in fact a misnomer: although electrolyte cannot escape from the battery, any gas evolved can. A more accurate description is 'valve regulated'. Therefore they may also be called valve regulated lead acid (VRLA).

Construction

The battery comprises positive and negative electrodes made of a mixture of lead and lead oxide with a microporous glass fibre separator and an electrolyte of dilute sulphuric acid. The electrolyte is held immobile in the absorbent separator material. The first SLA batteries used a gel additive to immobilize the electrolyte, but as technology developed, non-gel batteries have become predominant.

When a lead–acid battery is charged, a chemical reaction occurs between the electrodes and the electrolyte, leading to a potential difference between the positive and negative electrodes. During discharge, this potential difference allows a current to flow. The lead–acid battery is classified as secondary because energy must be applied before it can be used.

Batteries are built up on a 2 V cell basis to give 2–4–6 and 12 V blocks. In each cell of an SLA battery is a resealing vent which allows gas to escape in the event of overcharge but which prevents electrolyte escaping under normal conditions. With the vent correctly in place, a vacuum is created in the cell and a slight concavity may be seen in the walls of the battery. It is not possible to add water or electrolyte to a sealed battery, and it is vital to retain water in the system. The gases generated during overcharging are recombined so that they are not lost. Should oxygen and hydrogen escape from the battery, a gradual drying out would occur, eventually reducing capacity and shortening life.

Rating

Batteries are rated in terms of their voltage and ampere-hour capacity. The open-circuit voltage of any fully charged lead–acid cell is just over 2.1 V, so a 6 V battery with three cells in series would have an open-circuit voltage of 6.3 V, and a 12 V six cell battery would have a voltage of 12.6 V.

Capacity is more complex. It is expressed in ampere hours (Ah), and is the total amount of electrical energy available from a fully charged cell. However, the actual amount of available energy is dependent on the discharge current, the temperature, the end or cut-off voltage and general history and condition of the battery. SLA batteries are rated in the security industry on the basis of a 20 hour constant current discharge at 20°C to a cut-off voltage of 1.72 V per cell. As an example, a 12 V 6.5 Ah battery can discharge 325 mA (1/20 of 6.5 A) for 20 hours before the voltage drops to 10.32 V (6 × 1:72 V). The same battery will not, however, deliver 6.5 Ah for 1 hour (in fact, it will last about 30 minutes). When a battery discharges at a constant rate, its capacity changes according to the amperage load. Capacity increases when the discharge current is less than the 20 hour rate and decreases when the current is higher.

Charging

Constant charging is required for SLA batteries. The first requirement of charging is to apply a voltage at the terminals greater than the voltage of the discharged battery. In a discharged state the battery voltage is lowered, and is therefore most receptive to accepting current. If the charger output is great enough, the battery will receive current faster than its capability to use it in the recharge chemical reaction. The result is that the excess charge current converts to heat which, if prolonged, will damage the battery. In cases where a charging circuit can put more than 25–30 per cent of the rated capacity into the battery, then some form of current limiting is desirable. As an example, with a 6.5 Ah battery the limiting current should be no more than 30 per cent of 6.5 A, or 1.95 A.

In float use in the intruder alarm industry, the peak voltage output from the charging circuit should not exceed 2.25–2.3 V per cell or 13.8 V for a 12 V six cell battery.

The battery must be able to support the system to maintain continuous operation for a specified minimum period in the event of mains failure. The standby battery must also be automatically introduced on mains failure or if the mains voltage falls to a level below which the output voltage of the panel would not be sufficient to run the system. On 230 V supplies this equates to 216 V AC. Rechargeable batteries must also have sufficient capacity to run for a minimum period of 4 hours in alarm. On installation, the battery or batteries must be marked with the date. If ever the voltage goes low, an alarm must be triggered, and it should not be possible to set the system with low voltage.

Storage and self-discharge

Batteries self-discharge. It is only possible to devise battery chemistry systems where electrical leakage is minimized but not eliminated. Batteries are active, not passive, so chemical reactions always take place, which reduces the amount of available energy in the battery. In general, the rate of self-discharge is about 3.5 per cent per month, but this is accelerated as the storage temperature rises and decreases as it drops. For this reason it is vital that batteries be date coded and stored in a cool area.

Definitions

Before we can consider the selection of the battery there is a need to appreciate SLA battery terminology in so far as it relates to the intruder alarm area:

- *Ampere-hour* (Ah). The product of current in amperes and time in hours. It is used to indicate the battery capacity.
- *Capacity* (C). The rated capacity is the discharge capacity that the manufacturer of a battery claims may be obtained at a given discharge rate

and temperature. The available capacity refers to the electrical charge in ampere-hours that can be discharged from a battery based on its state of charge, rate of discharge, ambient temperature and specified cut-off voltage.

- *Float service*. This is a method in which the battery and load are in parallel to a float charger or rectifier so that constant voltage is applied to the battery to keep it fully charged and to supply power to the load without interruption or load variation.
- *Cut-off voltage*. The final voltage of a battery at the end of discharge or recharge.
- *Impedance*. The resistive value in ohms of a battery to AC.
- *Nominal capacity*. The nominal value of rated voltage/capacity. For intruder alarm purposes, this is equated to a 20 hour rate.
- *Open-circuit voltage*. The voltage of a battery when it is isolated from the load.

Discharging

With reference to Figure 4.1, and using a 20 hour rate capacity, by plotting discharge time against discharge current, the minimum capacity of a battery, expressed in ampere-hours can be found.

During discharge the voltage will decrease although the voltage will tend to remain high and almost constant for a relatively long period before declining to an end voltage.

When it is necessary to check batteries, initially it should not be disconnected from the power supply before examining and checking the float voltage. It is the float voltage that indicates how the battery has been treated since last examined, and once disconnected any anomalies owing to the float voltage affecting the battery will disappear.

A discharge test can then be used to prove the battery capacity. A high dummy load can be applied over a 10 minute period (Figure 4.2). This shows the variation of battery voltage with discharge rate. An example of a discharge test for a 3.0 Ah battery is to use a load constructed of two 10 Ω resistors in parallel. These will become hot, so should be rated for at least 25 W and mounted to dissipate heat.

Maintenance

A 5 year float life can be assumed, but in order that this is achieved, the battery must be operated under optimum float charge conditions. Prior to installation the engineer must:

- check for transit damage, ensuring that there are no cracks in the case or other external damage;

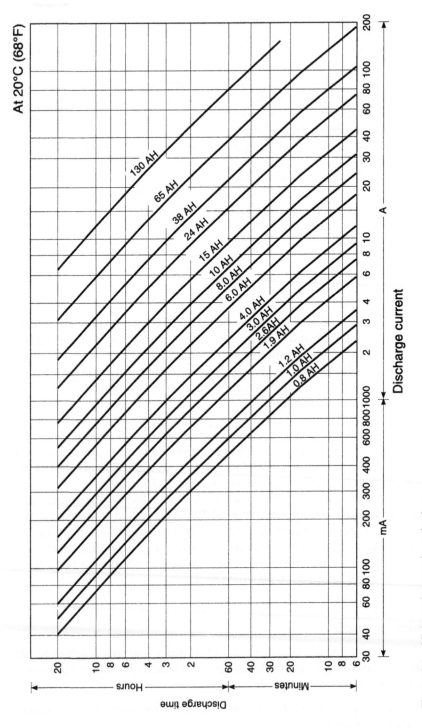

Figure 4.1 *Minimum battery size selection*

Figure 4.2   *Variation of battery voltage with discharge rate*

- ensure correct connection, and with the power supply on check the float voltage;
- give a period of charge and then conduct a short discharge test;
- record the installation date and float voltage.

Considerations and safety precautions

- Lead–acid batteries are damaged in terms of capacity and service life if fully discharged. If taken to zero volts and left off or on load, the internal battery resistance would elevate and the battery no longer accept charge.
- When a battery is operated in a confined space, adequate ventilation should be provided.
- Do not place a battery in contact with organic solvents or adhesive materials. When cleaning the case do not employ organic solvents such as petrol or paint thinners but use a water-soaked cloth.
- Use the correct terminals when connecting wires and avoid soldering.
- Avoid use outside the temperature range of –15 to 50°C for float/standby applications.
- Fasten the battery securely if it can be subjected to shock or vibration.
- Provide free space between each battery of 5–10 mm.
- If two or more battery groups are to be used in parallel they must be connected to the load through lengths of wires that have the same loop line resistance as each other – this ensures that every parallel bank presents

the same impedance to the load and thus correct equalization of the source to allow for maximum energy transfer to the load.

- Realize that the ripple current flowing in the battery from any source should not exceed 0.1 C amp RMS.
- Batteries should not be stored in a discharged state or be used alongside those of a different capacity, be of a different age or have been subject to a different use.
- Never try to dismantle or incinerate a battery as it can rupture.
- If any accidental skin/eye contact is made with the electrolyte, wash the affected area immediately with clean, fresh water. Immediate medical attention should be sought in the case of eye contact.
- Never install a battery close to equipment which can produce electrical sparks since the battery may generate ignitable gas.

4.3 Power supply requirements

The requirements in the European Standards are more stringent than those that had applied in the original BS 4737 which stated that a system must have an 8 hour standby capacity. The European Standards require a PSU to be capable of providing 1 A continuously whilst at the same time recharging a flat battery when the mains supply voltage is reduced by 10 per cent.

The batteries within the power supply are installed to ensure that the system remains stable in all conditions and not just in the event of a power failure.

The power supplies can be located within the system components (i.e. in the control unit) or in separate housings. Therefore the power supplies can be fitted as a single power source or multiple power supplies located around the premises. The same rules apply no matter which method is chosen.

There are three classifications for power supplies under the EN standards although it is Type A in which the intruder alarm surveyor has the main interest as it is the most flexible in terms of operational needs:

- **Type A.** The conventional type. A prime source and an alternative power source in the form of standby batteries which are automatically recharged by the intruder alarm system.
- **Type B.** A prime source and an alternative power source which is not replenished by the intruder alarm system.
- **Type C.** A prime source with a finite capacity and not recharged by the intruder alarm system.

EN50131-6 sets the battery back-up time in regard to the system grade. For intruder alarm systems this is referenced to Type A.

There are different requirements for the various system grades:

Duration of the alternative power source. Type A
- Grade 1 and Grade 2. 12 hours standby capacity.
- Grade 3 and Grade 4. 60 hours standby capacity.

Although Grades 3 and 4 are quoted as 60 hours, this may be reduced by 50% if the system is capable of signalling a mains failure to an Alarm Receiving Centre.

If a supplementary prime power source with automatic changeover between the prime power source and the supplementary prime power source is employed on a Grade 2, 3 or 4 system, the required period for the alternative power source to power the system may be reduced to 4 hours.

There are also special requirements for the recharge time:

Recharge time to 80% capacity
- Grade 1 and Grade 2. 72 hours.
- Grade 3 and Grade 4. 24 hours.

Although rather more specialist, the requirements for Type B and Type C power supplies need to be appreciated:

- **Type B.** Capable of powering the system for 24 hours for Grades 1 and 2
  Capable of powering the system for 120 hours for Grades 3 and 4
- **Type C.** Capable of powering the system for 720 hours for Grades 1 and 2
  Capable of powering the system for 720 hours for Grades 3 and 4

In basic terms it is possible to roughly estimate the required size of a standby battery although it is to be accepted that the true standby capacity is more complex. As an example, if all of the system loads can be added together and determined to continuously draw 600 mA then a Grade 2 system would in simple terms need a battery size in excess of 7.2 Ah from the formula:

600 mA $\times$ 12 hours = 7,200 mA or 7.2 A.

Therefore a battery in terms of Ah capacity above this value must be selected. However, there is a need to also take into account the current available for battery charging. Therefore consider a simple Grade 2 system with a quoted recharge time of 72 hours.

Consider a control panel with a 1 A power supply.
  Typical system quiescent current consumption consists of the following:

Control panel plus keypad   150 mA
External sounder   50 mA
Detectors   160 mA (8 × 20)
Total current   **360 mA**
Typical control panel   90 mA available for battery charging

Using a 7 Ah battery it will recharge in 72 hours and theoretically provide 910 mA (1000−90 mA) for the system. The supply rating for that system is 7.0/12 hours (standby capacity) = **583 mA** which is satisfactory to supply that system.

In addition to selecting a PSU Grade and Type there is a need to select a product classified according to its Environmental Class numbered I, II, III or IV listed in a progressively more severe or harsh environment order. The power supply must be classified to meet, or better, the environmental criteria for the location in which it is to be sited. These Classes are given in Section 1.2 of this book.

4.4   Ancillary duties

By far the most popular and efficient method of powering intruder alarm systems is via a power supply or UPS plus rechargeable secondary cells. Certainly the SLA battery is the most widely adopted type and little now is seen of the nickel–cadmium (NiCad) sealed battery, which is only generally used as a small cell for self-activating bell (SAB) modules or in some domestic dialling units as a standby source. NiCad cells are robust and dry with a high-temperature application use, plus a long shelf life; however, they are expensive compared with the SLA battery, and they have a lower voltage per cell. Leclanché cells of carbon–zinc construction may be encountered as a primary cell, and so also may lithium–manganese dioxide batteries. Leclanché cells and lithium–manganese dioxide, when used in systems that rely on primary cells only, must have sufficient capacity to support the quiescent state system load for the total period that they are to be installed plus 4 hours in an alarm condition. Batteries must be replaced at intervals not exceeding 75 per cent of their quoted shelf life, and the date of installation must be indelibly marked on the battery banks. Primary cells are non-rechargeable because they can be discharged only once. With secondary cells the underlying chemical reaction is reversible so that charging and discharge may be repeated many times.

A number of cells will be needed to make up the power source. It is important not to mix up batteries of different types and from different manufacturers. Connecting a group of batteries or cells by linking all the terminals of the same polarity will increase the capacity of the battery group without any influence on the voltage. By linking the terminals of opposite polarity an increase will only be attained in the voltage. These are parallel and series connection methods, respectively.

The need for additional electrical power, no matter how derived, as systems are extended, or in order to support auxiliary equipment cannot be underestimated.

From this the student can easily visualize many applications, including the driving of additional powered detectors, remote signalling equipment and zone omit units or such via the regulated auxiliary supply. These power supplies can also be used to boost the branch voltage on multiplex networks by installing the supplies at a number of points on the line. Using electromagnetic relays, the control panel trigger can switch the relay coil so that the power supply can generate extra potential to drive more or heavy duty audible or visual signalling devices.

Only BT-approved power supply units can be connected as an interface with the BT network. These power supplies are type approved as a result of the product meeting the test and safety criteria deemed necessary of goods that are to be connected to the BT lines. The BT-approved unit will be found to have additional circuitry compared with that of the standard device, with filters to give a particularly smooth output with sufficient regulation and ripple noise suppression for voice frequency circuits.

4.5 Power supply considerations

The advent of the stringent requirements of PD6662 and EN50131 led to the introduction of technically advanced power supplies with deep discharge protection, intelligent active battery monitoring and low quiescent battery standby current. In addition, features such as remote fault signalling outputs via volt free contacts for external power source (EPS), alternative power source (APS) or general fault became more commonplace. Remote maintenance became available on Grade 3 units.

The power supply has always been a vital part of the intruder alarm system as it supplies the power to the control equipment. This in turn controls the detection circuits and warning devices and is used to set and unset the system, to trigger outputs and remote signalling components. It will also provide indication as to the status of the system and give information regarding events.

This all meant that sufficient power must be made available by the PSU to drive the entire network plus any separately housed signalling equipment that does not have a unique power source.

Using proprietary measurement records forms it is useful to detail the battery location, charge voltage, current drawn and system standby capacity during the commissioning as this is an essential part of the intruder alarm network:

PSU Measurement records	
Battery location	*Remote psu. First floor office cupboard*
Battery charge voltage	*13.67 v*
Current drawn (full load of system)	*360 mA quiescent*
Standby capacity of system. Hours / minutes.	*13 h 30 mins. Grade 2*
Client. *AE Cont. Sys. Greenways*	

In general the protection of the power supply is by a fuse common for all circuits or thermistors that go to an extremely high resistance if an excess current creates heat in the resistor. Crowbar circuits are used if an overload current exists at the power source.

The standby battery will be separately fused as is the auxiliary power output.

Therefore we must consider the power supply as the energy source within the control panel plus any remotely located unit.

In practice we have mains electricity plus battery standby support to carry out the function with the unit British Telecom approved if remote signalling is to be incorporated.

In practical terms we need to envisage the power supply as being made up from a primary and secondary supply. With regard to the standby batteries it must be remembered that lead–acid cells are damaged in terms of capacity and service life if fully discharged. If taken to zero volts and left off or on load, the battery resistance will rise and the battery would no longer be capable of being recharged. SLA batteries in any control equipment or PSU are to be changed after a period of 4 years. The standby battery must be automatically introduced on mains failure or if the mains voltage falls below that which the output voltage of the control panel would not be sufficient to run the system – normally 216 Vac.

It can be noted that additional power supplies can be added to any system to improve the system security with diodes installed so that in the event of one PSU failure the current cannot flow back between the different units.

Fuses play an important role inside a power supply. Therefore it is always necessary to replace the blown fuse with the correct size and type which are generally:-

Mains input fuse – type **T (time lag)**
Battery and loads – type **F (fast acting)**

With all this in mind the principal selection considerations for a standard Type A power supply, taking into account that there is always a possibility of future system expansion, are:

| **Principal considerations of EN Standards** |

System Grade Calculate total system capacity
Determine standby battery

System Environmental Class ⇨ Verify environmental conditions

| **Further considerations** |

Need for future ancillary components
Integration with other systems/services
Future system extension
Additional PSUs to improve system security
If mains failure signalling to ARC a response plan must be in place

5 Intruder alarm control equipment

These days control panels used for intruder functions range from simple key-operated conventional systems to highly advanced premier control systems with multi uses. The technicians of today have to be aware of the security aspects of facilities available in order to maximize the potential of these systems and to minimize the risk to customers. Features such as tag setting and control from a keypad or reader are becoming more mainstream. We are also seeing great advances in the transmission of short message service texts to mobile phones plus a facility for maintenance to be carried out from a remote point plus internet-working or internet protocol (IP) facilities with the system being connected to the full range of internet features. The IP networking protocol also allows transmission of data so that it is possible to have cross system compatibility and huge levels of overall site control carried out via security interfaces.

Control panels will be found to have a selection of facilities extending from zone and anti-tamper features to bell functions including cut offs, delay, inhibit and test. The system facilities will also include isolation features and resetting methods with options for setting and unsetting with appropriate time modes. All these facilities need be understood alongside methods of latching and freezing to aid identification of detection devices that have provided signals at a particular point in time.

In addition to the system facilities, there are many detector circuit facilities available to provide flexibility. These attributes will range from the way a circuit can be programmed for operation through to methods that can be used to include anti-false alarm features and night alarm and test facilities.

These facilities themselves become more complex as we move from conventional control panels to multiplex panels that have all the detection circuits wired to the principal circuit board via a data collection network.

All of the control panel terminal descriptions and locations plus the inputs that may be applied need to be understood. These can relate to single- or double-pole or end-of-line configured devices and addressable points.

In the section on common forms of outputs the alarm triggers and bell relay contacts and options on signalling will be covered.

When discussing control panels ancillary control equipment must also be considered, and processing devices and zone detection circuit indication panels understood. There is also a requirement to extend knowledge to digital keypads, pass switches, shunt locks and zone detection circuit omit devices.

With the progression in technology, remote interrogation is becoming increasingly used and alterations to programmed systems via uploading and downloading protocols made. Additionally, the problems caused by induced electromagnetic energy in signalling lines and its suppression need to be understood.

The intruder alarm equipment can therefore be seen in the wider scheme as a complete flexible family but with consistent hardware and software.

5.1 Control panel system facilities

The need for a control panel to be simple to use, flexible, secure and reliable in operation cannot be too highly emphasized. The operation of the system occurs in three phases. These are energizing of the system, fault testing and arming.

The energizing of the system or the 'turning-on' procedure now tends to involve the entering of a unique code into a keypad rather than the original method of using a key lock. Following this procedure the control panel will fault test and identify any circuit that is not clear and needs attention. The arming process can then commence through an arranged setting method. This will involve either a timed exit, in that the system will be set after a preprogrammed exit timer has expired or after a final exit door has been opened and closed. Alternatively it can be set after an exit terminator button on the outside of the premises has been pressed, or the setting may be instantaneous if the premises are not to be vacated. However, the system must be unable to arm in the event of a sensor being at fault. When remote signalling is applied, the ARC can monitor for setting at prearranged closing times to ensure that the correct procedures have been followed.

The introduction of EN50131 brought about many changes relating to the actual end user operation of the control panel and the way that the security system was to be operated. Even though many new requirements have been introduced over the years, the intruder alarm panel, in broad terms, still detects and signals.

We also know that in accordance with the EN Standards that compliance plays a major part when it comes to the use of the control panel. Therefore the manufacturer has a great deal of responsibility in the design of the equipment to meet a certain security grade and to engineer it so as to ensure it may be used with a range of accessories in and in a variety of applications.

We will certainly see at the top end a product that will typically be marketed as:

> **EN50131/PD6662: 2004 Grade 3 compliant**
> **264 zones. iD circuits. Radio links**

Upload/download control panels that claim to be in the same price parish as 'bells only' panels have also become widely accepted. These have integrated communicator and modems with PC software and can fit into both the domestic and light commercial sector. They enable certain system parameters and settings to be amended and managed from a remote PC.

A great deal of work was in fact required by the equipment manufacturers to bring their alarm panels or control and indicating equipment (CIE) in line with PD6662: 2004. Completing detailed product specifications and re-writing software was a major engineering programme – indeed in some instances was akin to developing a new product.

In reality the real parameters built into the control equipment are governed by the equipment manufacturers in relation to the grade that the product claims compliance with via an official Certificate of Conformity. In all cases the installer will be given a range of options with the different equipment. However, certain functions are mandatory. For example, the higher the security grade of the system the more digits are required in the setting codes for Access Levels 2, 3 and 4. See Section 1.2.

For Access Levels 2, 3 and 4 the following apply:

- **Grade 1 system codes.** These may be 3 or more digits
- **Grade 2 system codes.** These may be 4, 5 or 6 digits
- **Grade 3 system codes.** These must be 6 digits
- **Grade 4 system codes.** These must be 6 or more digits

For mainstream applications we are mostly concerned with Grade 2 and 3 systems. However, with equipment at even the lowest Grade 1 the equipment log must indicate system activity and the cause of alarms etc.

As we graduate to Grade 2 and upwards, other important issues that become more clearly defined than those that were employed in traditional old-generation systems are those of display blanking, system setting/ unsetting, faults during Day (unset conditions) and engineer access and the way that these conditions must be notified to the end user.

Under the European Standards a lot of emphasis was put on the control equipment (or user interface), the alert displays and the hiding of information.

It is to be emphasized that all mandatory indications have to be located in one position; additional indications can however be located at other locations throughout the supervised premises.

Although such particular display parameters are the responsibility of the control equipment manufacturer, there is a need for the installer to understand their significance.

Display blanking

It is allowed for the system display to show the status (i.e. system set or unset) for a short period of time. However, it is not allowed that the system permanently displays the system activity.

Depending on the system grade when the setting has been completed an indication is required which is time limited to a maximum 180 seconds to annunciate that the system is set or part set. Therefore once the system setting has been completed it is not obvious as to whether the alarm is actually set or unset.

Therefore as the system sets or unsets it will indicate as an example:

```
<<  SET 1  >>
01 Feb  17:12:48
```

or

```
<<  UN-SET 1  >>
01 Feb  17:12:48
```

For ten seconds and then display:

```
01 Feb  17:12:58
```

However it is always possible to view the current status of the system by entering the user code.

System setting/unsetting

In the event that a non entry/exit zone is activated during the setting procedure the display as an example must show:

```
<<  Can't Set >>
   Zone 4 fault
```

If the zone faults are then cleared the setting procedure can continue.

Setting may be initiated within the supervised premises and completed outside of the premises or completed outside of the supervised premises using appropriate ancillary control equipment.

If the system is not in a normal condition when setting e.g. all detectors are closed, the system should not be allowed to set and only exceptional circumstances allow this requirement to be overridden.

Depending on the system grade, once the setting procedure has been completed, a system indication must be made to confirm the actual type of setting that has been completed such as full or partial setting. This has a time limitation of 180 seconds maximum. We also know that the exact type of system set function is not displayed.

Where the setting procedure has been satisfactorily completed there should be an indication that the intruder alarm system is set, except for Grades 1 and 2 of EN50131 where indication of set is optional. The set indication is always to be time limited and indicated to the user at the time the system becomes set. Consideration should also be given to the provision of extra internal audible indications prior to setting. This will alert people in the building that the system is due to be set.

For systems that include an entry route only 45 seconds is allowed to complete unsetting so as to ensure that entry routes are kept as short as possible. If this time is exceeded an alarm output must be generated. Equally if an alarm occurs during the unsetting it must signal an internal warning device. The completion of unsetting must of itself be displayed for 30 seconds maximum. If remote indication is used the alarm is not to provide remote indication until the internal warning device has been operated for 30 seconds maximum.

Faults during Day (unset)

If a fault occurs and the system is in the Day (unset) state, the fault sounder is to operate and the display is to show the problem such as:

```
!! Check !!
```

In order to establish the cause of the fault the user code must be entered. The display will then indicate the source of the problem. As an example if the problem is a bell box tamper:

```
┌─────────────────────┐
│    !!  Check  !!     │
│   Bell box tamper    │
└─────────────────────┘
```

Such a message must stay on the display for some 10 seconds or if an attempt to reset the system is made it will then once again return to **!! Check !!**.

Once the fault has been rectified the system needs to be reset and at this stage the screen will return to its normal standby mode.

Engineer access

The user must authorize engineer access. When the user code is entered it is necessary to scroll through the options until the following example is reached:

```
┌─────────────────────┐
│      <Authorise      │
│  engineer access ? > │
└─────────────────────┘
```

If the Yes (confirmation) key is selected the display will then show the level of engineer authority such as 'Master'.

```
┌─────────────────────┐
│  << Engr authorisor  │
│     = Master >>      │
└─────────────────────┘
```

The engineer can then gain entry to the system in the normal way by entering a valid engineer access code.

Other control panel system facilities of particular significance are the restoring of systems (resetting), indications and inhibit/isolation operations.

Restoring of systems (resetting)

In so far as alarm conditions are concerned user reset is normal and remote reset can be applied at access level 2 (normal/standard level of user codes) if information is available to determine exactly what has been the reason for the alarm condition.

Resetting of tamper conditions is rather different and although at Grades 1 and 2 resetting of tamper conditions can be made by a user with access level 2 this is not possible at Grades 3 and 4 as these must be engineer reset.

Indications

All mandatory system indications are to be located at one designated position. However, further indications may be located at additional points within the protected premises.

A maximum of ten switch contacts for doors or windows may be indicated on a common zone but energized detectors such as movement detectors are to be individually identified.

Inhibit/isolation

Only access level 2 users can inhibit system parts to inhibit system zones etc. When the system is unset the inhibit function must be cancelled to ensure that it cannot be restored accidentally when the system is next set.

Isolation of the actual functions that the system is capable of doing is different. It varies in regard to the system grade. At Grades 1 and 2 it is possible for the normal user to amend many system functions, but for the higher grade systems it must be carried out by the installer/engineer code.

Siting of control equipment

In all installations the control indicating equipment (CIE) must be sited within the supervised area. If systems are divided into subsystems of various grades the CIE must be located in the highest graded subsystem.

Therefore the considerations in regard to the siting are:

● Location
● Ease of operation
● Prevention of observation by unauthorized persons as to system codes

The control panel system facilities differ from system to system and will vary between control panels in the way that energizing, fault testing and arming is performed, but the general concepts can be summarized.

Single zone

This is a closed-circuit loop for connection to alarm detectors. In practice, it operates as an area protected by a detector. The single zone is common to all units as the principal mode of operation, and may be a simple loop with a separate anti-tamper circuit or have end-of-line resistors or use a multiplex or ID (intelligent device) method.

The single zone can be split using an annunciator to enable identification of tripped sensors with an existing single-zone unit by employing resistors of different value in conjunction with one line, but the system cannot obviously be controlled as efficiently as a multizone system. When more than one detector is used on a zone it should be arranged to latch indicator LEDs on zone splitters or detectors to allow identification of the unit tripped.

Zones may be configured in many ways as attributes.

Multizones are selected to suit the application, and allow future expansion to ensure total flexibility.

Anti-tamper (A/T)

Anti-tamper is a facility for all the wiring zones and signalling equipment, and gives the wiring 24 hour protection from damage or unauthorized interference of equipment.

Activation of the tamper circuit will provide audible warning and/or indication. The detection of tampering during the day may only cause triggering of internal sounders. To operate the full alarm, the facility can be differentiated on 'set'.

Certain control panels will feature global tamper covering of more than one zone or loop.

Bell cut off

This facility allows intelligent automatic resetting and rearming.

At the conclusion of the bell output that has been programmed, visual signalling can be continued. If circuits are clear then they may be rearmed.

With programmable systems it should be understood that if zero bell duration is selected the sounder will not operate when an alarm condition occurs.

Bell delay

This is a delay between a detector being activated and the panel triggering the sounder. It is used with remote signalling to send an advance silent signal.

If infinite delay is selected the sounder will not operate when an alarm condition occurs.

Bell test

This allows testing of signalling devices with control equipment in the day condition and avoids the need to vacate premises and set detection devices. The sequence may be:

- bell sounder operates;
  ↓
- extension speakers operate;
  ↓
- strobe and visual devices operate;
  ↓
- keypad response;
  ↓
- additional outputs respond.

Circuit isolation

When a zone is programmed for 'omit allow' the system will allow a zone to be omitted by the end user if required. This must be performed within a given period of time and be indicated to the user.

On unsetting this may be indicated to the user to show how the system was last set and to what level.

Engineer reset

This ensures that data cannot be lost by control panels that will not store information once unset and further reset.

After a user silences a system, a reset cannot be done until verified by the engineer or it is reset from a remote point to an agreed procedure.

Remote reset

Remote reset when enabled allows a user to reset the system under alarm company supervision over the telephone without calling out an engineer. This ensures that the user contacts the ARC to acknowledge accidental alarms or system problems.

To allow the customer to use the remote reset facility the control panel is programmed for engineer reset, and a central station identification code (CSID) is installed.

The remote reset is carried out by the user following a special sequence after an alarm activation. A number generated by a keypad display is passed verbally to the alarm company central station, and following a security verification procedure involving a code name or number the number is entered into a programmer which generates a reset code. The user is given this code in order to reset the control panel.

First zone to alarm

After an alarm activation the zones that tripped can be displayed in sequence. This allows the route taken through a premises by an intruder to be established.

Lock out

If programmed for first zone lock out the first zone to initiate an alarm condition will be automatically locked out and will not be included when the system automatically rearms after the bell cut-off period has expired. This is an option for automatic rearm.

Automatic rearm

If so programmed, when the bell time expires then all zones that are closed will rearm, and even the first to alarm will be reset for the programmed number of resets.

Night/partial set

In normal alarm (night) mode any activation will cause an instantaneous alarm. Partial set allows zones not programmed to be armed when part set is selected to be omitted.

Areas making up a number of zones can be pre-determined and selected so they can be omitted at the control panel using a more simple setting sequence.

Setting on final exit

This can be achieved in a number of modes.

Timed

A programmed period of time is allowed between the point of switching to set and actually exiting the protected premises.

Final door

The system will set some 3 seconds after the final door contact has changed state from open to closed. The exit time must be set to infinite or the system will set on expiry of the exit time.

Lock set

The system will arm after the final door contact has changed state and the switched lock has changed from the closed to the open state. The exit time will automatically set at infinity. On entry, the lock terminals must be closed before the final door is opened.

This function is used with an arming key switch.

Exit terminate

The system will arm when the exit terminate button is pulse closed. This will terminate any remaining exit time, which must be set to infinity. This will ensure that an exit fault does not occur by exceeding the exit time before pressing the terminate button.

Deferred set

When the system is armed the exit time starts with the programmed exit time. If a settable zone is violated on exit the timer restarts after a period of the order of 20 seconds and then starts counting down again. If no violations occur during this time the system will set on expiry of the timer.

Each time a zone is violated the timer restarts after a time interval of the order of 20 seconds.

Instant set

If there are no faults present the system will set instantly and silently. If a fault does exist the keypad will emit an error tone and cycle through any faults for the duration of the programmed exit time before returning to the day or unset condition.

The exit time is then programmed for a given time or is infinite for lock set or exit terminate.

Timed entry

This facility allows the final exit detector to start the entry timer. The system is unset by entering the code or turning off within this period. The entry period may be subdivided, and separate alarm tones give an indication of the entry period almost being exceeded. This constantly reminds the user of the urgency to disarm the system.

With lock set, the lock terminals must be closed before the final door is opened.

Entry abort

Entry abort reduces false alarms caused by user errors on entering a premises. This feature delays signalling the ARC policing an alarm signal by 90 seconds, usually during office hours (7.30 a.m. to 8.30 p.m. Monday to Friday). Providing the alarm receiving centre can accept 'auto abort' and 'entry abort', this is achieved by the control panel sending a 'restore' signal if the user exceeds the entry time and then enters a valid access code within 90 seconds after the alarm has been generated.

The system must then be reset by the engineer or by following the 'remote reset' procedure.

Walk test

This facility allows the user or engineer to test all alarm devices without the need to actually set the system.

It is usually performed by entering a programme at the panel, which will then give an audible tone and indication every time a circuit is opened:

Walk test
↓
System automatically displays circuit violations or system faults
↓
Internal sounder activates when circuits are open
↓
Cancel

Zone indication only

This disables the zone for a period of the order of 14 days. During this time if the zone is violated the event is logged only and no signalling is activated. If at the end of the period no violations have occurred, the circuit is automatically restored as a normal zone.

Latch

This switches after the exit time and/or a walk test and resets on disarming or the start of the entry time.

To identify a detector that causes an alarm when a zone has more than one movement detector or powered device the programmable output of the set latch should be wired to the latch input of all the detectors in that zone. When a detector causes an alarm the panel will deactivate the output – latching the LED on that detector and inhibiting all others. The panel also deactivates the output in day mode (Figure 5.1).

Latch freeze test

First to alarm (FTA) with control (freeze) and test (LED) outputs can be accommodated by certain detectors (Figure 5.2). This is achieved by selecting 12 V to be applied to terminal 1 and 0 V to terminal 2 when the panel is set. The latch is then set with an alarm and the control line drops to 0 V. When the panel is unset, test goes to 12 V to enable the LEDs on units that caused the alarm – the first flashing and the subsequent ones continuous.

The control panel system facilities are complemented by the detector circuit attributes, which are considered in the next section.

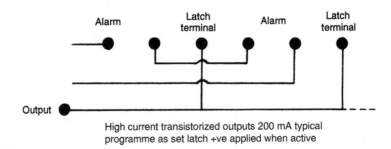

High current transistorized outputs 200 mA typical
programme as set latch +ve applied when active

Figure 5.1 *Latch facility*

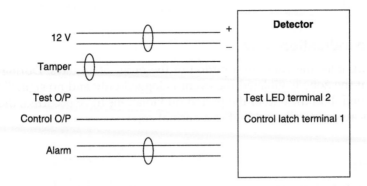

Figure 5.2 *Latch freeze test*

5.2 Control panel detector circuit facilities: control unit features

All detection devices, no matter how complex or simple, are connected to detection circuits. Each circuit should then be allocated a number which identifies the detection device. As an example, a room protected by a movement sensor may be circuit 01 while a door protected by a magnetic contact may be circuit 02.

There are attributes allocated to these detection circuits as follows:

- *Personal attack (PA)*. These may also be called panic alarms and are hand or foot operated to activate an alarm. This alarm signal will also be transmitted to the central station if remote signalling is employed. The ENs refer to them as hold up (HU) devices. The PA is constantly monitored to give a signal at all times on its own channel when used in conjunction with a remote signal. It can be programmed for an audible (bell output) or quiet signal. Duress signalling is a means of entering a passcode into a control panel to generate a silent alarm via the digital communicator to the central station.
- *24 hour*. A circuit that is monitored at all times is designated 24 hour. When triggered in the unset condition a local alarm of internal sounders is generated. When triggered in full or part set a full alarm is generated. Its output is therefore active every time a circuit programmed as a 24 hour type opens, and deactivates when the zone is closed and reset. Selection can change from a positive to an active loop for use on fire doors, monitored wiring and foil circuits.
- *Auxiliary*. The auxiliary circuit is one which is monitored at all times. When triggered it will activate outputs that have been programmed for special duties. It is not to be confused with the auxiliary supply circuit which provides the regulated supply for powered detectors.
- *Tamper*. This causes an alarm when the system is interfered with physically or there are too many unauthorized keypad entries. Its input is always operational, so if opened it will operate internal sounders in a disarmed state or full alarm when set.
- *Night circuit alarm*. This is defined as a detection circuit type that will generate a full alarm when the alarm system is set. When the alarm system is unset the detection circuit may be triggered without causing an alarm. The student should be aware that many domestic systems automatically deselect zones for night-time use or when selecting 'home', when the state of detectors will alter from alarm to access during the exit/entry time. This, however, is not to be confused with night circuits but is related to 'omit' or 'part set'.
- *Omit*. This allows the intentional exclusion of the monitoring of one or more detection circuits when full or part setting the alarm system.

- *Part set.* This is the state of the alarm system when it is protecting only a part of the premises.
- *Exit/entry.* This is the final route selected to leave and enter a protected premises. The final exit is the door or detector that is used for leaving and entering the premises and the exit/entry times are the preset delays to leave or enter the building when a timed method is specified.
- *Terminate.* The system sets when the terminate button is momentarily closed. It terminates any remaining exit time and confirms system set.
- *Chime.* This operates in the day state with activation of a buzzer or internal sounder only. It is an optional facility often employed on external shop doors to enable a selected detector to generate a two-tone chime specifically accommodated on many control panels.

A systematic attention to false alarm sources is also incorporated not only in the detector circuit but also in the control unit features. A feature called 'soak', when selected, will log the operation of a zone only if it activates when the system is set, and will activate only the service LED. It is intended to confirm the stability of detectors, as the data can then be stored in event logs, which are records of system activity held in non-volatile memories. The term 'paired' is applied to two zones that must both be opened after the alarm is set to generate an alarm, and signal processing or double knock is adopted to ensure that a number of activations are generated within a selected time window before a signal is accepted as an alarm condition. This is sometimes referred to in the detector itself as event verification, which in passive infrared devices (PIRs) intelligently analyse some four decision criteria such as target size, duration, repetition and time between repetitions. The term 'event verification' is often also used as a form of alarm confirmation which covers a far broader aspect of false alarm avoidance, and this is discussed separately in Section 6.9.

One of the main false alarm sources is actually during the opening and closing periods of premises, and this can be addressed in a few ways.

Remote signalling devices

With remote signalling devices, interfaces designed to interpret open and closed signals from the control panel can instruct the communication device to transmit particular instructions to the central station:

- *Exit/entry abort.* On system set a timer is started, and if during that time scale an alarm condition occurs, the interface will signal the communication device to transmit an intruder code to the central station. Should the user re-enter the premises and key in the passcode before the allowed time scale, then a restore signal or separate signal will be relayed to the central station.

- *Alarm/abort.* This is similar to 'exit/entry abort' except that there is no override on the bell delay.

Wiring is installed between the communication outputs on the control panel and the communication device. The input/output links are set for the correct polarity for the control panel being used. The bell delay can be overridden during the abort time by wiring the changeover contacts provided to trigger the outside sounder. If a separate code is required as an abort signal the 'ABT' output can be wired into a spare input on the communication device.

Remote signalling system tests use the following methods:

- *To test 'exit abort'.* Set the system and generate an alarm, then enter the user code within the allowed time scale. Check with the central station to ensure that it has received the code followed by 'restore' or 'abort'.
- *To test 'alarm/abort'.* As for 'exit abort' but wait beyond the allowed time scale.

Figure 5.3 shows the typical terminal layout for a stand-alone module that may be connected into an existing system.

Figure 5.3 *Terminal layout abort module*

Variants of the stand-alone module can be sourced for use with other control panels, and placed between the communication pins on the control panel and the communication device. They may also perform a sequential alarm confirmation task and control the outside sounder, strobe and internal sounders.

It is possible to send a confirmation alarm signal by means of a sequential alarm confirmation module, which if an alarm is generated will send an intruder signal (usually code 3). Once this has been transmitted to the central station the zone sending the alarm will be locked out. The module then scans the other zone inputs.

On activation of a further zone a confirmation signal (usually code 8) will be sent after the alarm abort (channel 7) time window. This informs the central station of a confirmed signal (see also Section 6.9). Figure 5.4 shows a typical jumper selection.

The jumpers in the stand-alone module are defined below:

- *Snd-Sel.* This is the sounder select link. The bell delay can be overridden within the abort time by leaving the link connected. In this mode an artificial 'comms fail' signal is sent to the control panel to override the bell delay. To differentiate this event from a genuine 'comms fail', two consecutive 'comms fail' entries are made on the control panel log.
- *Tell Back.* This allows the user to reset the control panel if an alarm on exit has been aborted. To disable this function, remove the link.
- *Test.* This disables test or enables factory test only.
- *Abrt Sel.* This is used if a separate code is required as an abort signal. The link is therefore left connected. In this mode a given channel (usually 5) of the communication device is triggered. The input will send a code 5 to the central station as an abort signal. If a different code is required, channel 5 should be programmed to send the necessary code.

Under the Association of Chief Police Officers (ACPO) and NACP 14/SCOP 103 covering National Approvals Council for Security Systems (NACOSS) and industry requirements, all remote signalling systems,

Jumper connnections		
Snd-Sel	- On = Use comm fail to control audible warning Off = Use relays to control audible warning	
Tell Back	- On = Allows user reset within Off = Disables user reset	Abort time
Test	- On = Disables test Off = Enables factory test only	
Abrt Sel	- On = Abort signal on Channel 5 Off = Disables Abort on Channel 5	

Figure 5.4 *Sequential alarm confirmation*

whether operating by telephone or radio, must meet the British and European Standards by incorporating user-generated abort signals or open and close signals. Although unsetting after an alarm has occurred when the system has been set is relatively straightforward, involving application of the user code and then logging of the recorded event, the method of remote resetting after unsetting can only be performed by a special procedure. A typical method is illustrated in Figure 5.5.

The student will be aware that the initial access to the system shown in Figure 5.5 was gained by entering a four-digit passcode. With most systems a number of separate user passcodes are available for operating the system. The master user is the person responsible for allocating the other users to the system, and these extra user codes may be defined for different authorization levels. These user codes will then be assigned to a user type (code level). Once the master user has assigned a user to the system,

Figure 5.5 *Unsetting after an alarm*

users will be allowed to change their passcode (personal identification number or PIN) but not the user type. Examples of typical user types or code levels are shown in Table 5.1.

Consider how the code levels in Table 5.1 can be configured:

- *Master*. Full access to all menus and options.
- *Standard*. Access only to menu 1.
- *Holiday*. Allows the system to be set and unset and access to menu 1. Can be assigned to a temporary user during holiday periods.
- *Set only*. Allows access to set the system and menu 1.
- *Reset only*. Allows 24 hour alarms to be reset and access to menu 1.
- *PA code*. Allows use of a silent duress signal or audible signal PA code.

It follows that a user with the master code level has full access to all options (Table 5.2). However, a user with a code allocated as 'holiday' has authorization to access menu 1 (certain options from this menu may not be allowed, depending on the specific equipment specification). Equally, there is no access to menu 2 or 3.

With respect to levels authorized by codes, a term that the engineer will sometimes encounter is 'cleaner'. This is to account for cleaners entering

Table 5.1   *Code level and access to menus*

Code level	User menu 1	User menu 2	User menu 3
Master	✓	✓	✓
Standard	✓		
Holiday	✓		
Set only	✓		
Reset only	✓		
PA code	✓		

Table 5.2   *User menus*

User menu 1	User menu 2	User menu 3
Bell test	View circuits	Part set groups
Walk test	Set clock	Code set groups
Remote reset	Set date	User's name
Change passcode	Set up users	View inactive circuits
Enable chime	Alter chime circuits	
Omit 24 hour group	Alter 24 hour group	
Omit circuits	Print system log	
Silent set	Configure wards	
View activity count	View system log	
Full set/unset	Remote call back	
Part set/unset	Remote service call	

commercial premises outside of working hours, and allows the cleaners access to certain parts of the premises to carry out their domestic duties.

The student may have also noted in user menu 2 that options for remote call back and remote service call are listed. The former covers systems fitted with a modem which enable the installation company to obtain data from the control panel. Initiating of remote service calls allows uploading from a remote site. The upload and download characteristics are considered in Section 5.7.

The programming of code levels allows a further function, 'wards', which are groups of detection circuits that may be set or unset independently of each other. Consider a typical three-bedroom house (Figure 5.6).

Figure 5.6 *Three-bedroom house protection*

Table 5.3   *Ward programming (domestic)*

Circuit no.	Location	Part set A	Part set B	Part set C
01	Front door	A	A	A
02	Smoke detector	A	A	A
03	Hall door to garage	A	A	A
04	Garage up and over door	A	A	A
05	Garage door (back)	A	A	A
06	Kitchen door	A	A	A
07	Dining room PIR	A	A	O
08	Lounge PIR	A	A	O
09	Bedroom 2 PIR	O	A	O
10	Landing PIR	O	O	O

The object is to define part set buttons to omit one or more wards. The easiest method of configuring the part set groups is to take the information from a table and translate it directly to the ward programming details. The part set arrangements required are:

- *Part set A*. Downstairs perimeter detection armed and downstairs internal detection armed.
- *Part set B*. Downstairs perimeter detection armed, downstairs internal detection armed and bedroom 2 armed.
- *Part set C*. Downstairs perimeter detection armed.

First a table (Table 5.3) listing the circuits that are required to be armed (A) and omitted (O) for each part set requirement is generated. From this, the detector circuit numbers are programmed into the part set functions so that during the setting procedure the user need only further select button A, B or C to arm the appropriate wards. This is displayed and identified on the control panel indicator. If no part set is selected, then all circuits are armed. If, however, as an example, button B is pressed, then only the detectors on part set B are armed (including C1–C9).

Although we have discussed a domestic application, the same principle can be extended to a commercial part set application. Figure 5.7 shows typical commercial premises. The requirements of this example application are:

- the sales manager requires independent access to the sales department and the canteen;
- the workshop manager requires independent access to the workshop area and the canteen;
- the stores manager requires independent access to the stores area and the canteen.

Figure 5.7   *Commercial premises protection*

From the above requirements each area can be assigned to a ward, e.g.:

- *Ward A*. Sales department and canteen.
- *Ward B*. Workshop and canteen.
- *Ward C*. Stores and canteen.

Using the above ward requirements, a table (Table 5.4) for assigning circuits to wards is generated. Part set is reversed.

The methods we have studied show great flexibility. There are in fact other systems that will be found using ancillary control equipment such

Table 5.4   *Ward programming (commercial)*

Circuit no.	Location	Ward A	Ward B	Ward C
01	Sales entrance door	A	O	O
02	Sales PIR	A	O	O
03	Workshop PIR (1)	O	A	O
04	Workshop entrance door	O	A	O
05	Workshop PIR (2)	O	A	O
06	Workshop internal door	O	A	O
07	Workshop PIR (3)	O	A	O
08	Sales internal door	A	O	O
09	Canteen door	A	A	A
10	Canteen PIR	A	A	A
11	Stores entrance door	O	O	A
12	Stores PIR	O	O	A
13	Stores internal door	O	O	A
14	Sales (office PIR)	A	O	O
15	Service entrance door	O	O	O
16	Service PIR	O	O	O

as shunts to provide limited access to controlled areas, and we consider these later in Section 5.4.

5.3   Conventional connection details

Before we look at the specific connection terminals on a control panel we should have some understanding of the circuit that is used in a simple door alarm to produce a latching condition once a detector switch has been operated (Figure 5.8).

The circuit is based on the action of two cross-coupled NAND gates which form a bistable flip–flop circuit. In the example, inputs at pins 2 and 13 are normally held at a logic high level. As the name 'bistable' suggests, the flip–flop is stable in two states. In its set state, pin 3 is high and pin 11 is low, whereas the converse applies in the reset state. Normally the flip–flop is in its reset state, and pins 2 and 13 are high. The output at pin 3 is low, and no current flows to the transistor (Q1). Since Q1 is off, no current can flow to the sounder.

Input at pin 2 is held high because the pin is connected to the positive rail through a normally closed switch (SW2). This is the detector contact in which we are interested. When this switch opens, the voltage at pin 2 falls, pulled down by the resistor (R1), setting the flip–flop with the output at

Figure 5.8 *Simple door alarm circuit*

pin 3 going high. This turns on the transistor, causing the sounder to operate with the alarm being controlled by the flip–flop. Even if the detector switch is closed, the sounder will not be silenced. To silence the alarm the flip–flop must be reset, which is done by pressing and closing switch SW3 to give a brief low input at pin 13.

We should also note that the circuit has a measure of anti-tamper protection, as cutting the wires between the switch and the rest of the circuit will have the same effect as triggering the detection switch.

Clearly some knowledge is needed of the fundamentals of the circuit employed in analysing and controlling the detectors, although it forms only part of the control panel circuit. Although control panels have various functions geared to the same ends, even the terminal layout will vary considerably between models. When considering control equipment, all detection circuits must latch and every zone give an audible or visual indication of the alarm condition existing during setting or unsetting or when testing the system. This allows manufacturers of control panels a high degree of latitude, and hence the electronics within different panels will vary enormously, and more so with the growing use of remote setting keypads. It is still important, however, not to select equipment that can overburden a system or overcomplicate its operation and generate an increased risk of operator error.

In the first instance we will look at a conventional control panel terminal layout intended for the domestic or small commercial sector (Figure 5.9). If we consider this panel in overview, it is a six-zone microprocessor based with five programmable zones plus final exit. The five programmable zones are 'night', 'access', 'fire', 'PA' and 'key-switch'. The tamper protection is

Figure 5.9 *Conventional terminal layout*

common or global, and covers the supervision of all the zones in one circuit or loop. The final exit zone is dedicated.

The control panel also has provision for connection of a number of remote keypads to the system to enable remote operation from different points in the protected premises. These remote keypads will have the same indications as the main panel and an internal sounder to indicate all system tones. The wiring of the remote keypads is achieved by plugging an interface socket into the main circuit board of the main panel and taking a multicore cable to the keypad, using a star or daisy chain configuration if more than one remote keypad is to be installed.

The wiring of the detectors is conventional with pressure mats wired across the tamper and zone terminals.

- ALM^-. This terminal is switched negative (100 mA) in alarm and is removed when the system is reset. It can also be disabled when the system is set to 'home' if a link on the circuit board is closed. This terminal and its output can be used to trigger such devices as speech diallers.
- RST^-. This terminal can be programmed in two different ways:
  (1) *DIR RST*. When programmed as such ('detector reset') the output may be used for detectors which latch their alarm condition, and must be depowered to reset (e.g. smoke detectors and vibration detectors). Power for such detectors must be connected between AUX12V+ and RST^-.
  (2) *SW12V*. When programmed in this way ('switched 12 V') the output may be connected to the latch input terminal on latching detectors such as PIRs to identify detectors that have alarmed when more than one are connected on a single zone. The LED that has latched will clear once the system is reset.
- TRG^-. This is a negative trigger in alarm. It is used as the sounder output but can be programmed as SAB or SCB. When selected as SAB the output will switch to 0 V on alarm, and hence should be connected to the negative trigger input on an SAB-type sounder module. If programmed as SCB the output provides a 0 V hold-off which is removed in alarm, and in this case must be wired to the negative trigger input on an SCB sounder module.
- STB^-. This is a strobe output providing a switched negative signal in alarm that can be programmed to continue after the sounder has timed out.
- H/O^+. This provides a permanent 12 V supply hold-off voltage for the sounder module and is generally protected by a 1 A bell fuse. It is for connection to the 12 V supply input on the sounder module.
- H/O^-. This is a permanent 0 V hold-off for the sounder module.
- TMP^-. This is the tamper protection for the sounder. It is connected to the negative tamper return output on the sounder module.
- *AUX TAMPER*. This is the global or common tamper protection.

- *L/S⁻*. In this case a 16 Ω extension speaker may be connected between the L/S⁻ and AUX12V+ terminals. The R37 component located in the corner of the circuit board can be cut to reduce the volume of the internal sounders although the alarm is always at full volume.
- *AUX12V*. These terminals provide the regulated supply for powered detectors which require a low-voltage supply and are protected by an auxiliary fuse, generally rated at 1 A.

It will be seen that zone 5 can be configured as either alarm or PA and zone 4 as either alarm or fire. When chosen as fire, in the event of activation, a distinctive internal sounder tone is generated and the external sounder is pulsed.

Zones 2 and 1 can be programmed as either alarm or access. These are normally ascribed to detectors between the control panel and the final exit. Access zones enable walk through during setting and unsetting but will provide a full, immediate alarm if tripped before the final exit detector.

If at a remote point the system is intended to be set or unset by a physical key in preference to a code, then a keyswitch can be connected in zone 3.

The final exit zone is simply the point at which the user leaves and enters the premises. When re-entering the premises the user must activate the final exit zone to start the entry time, which allows the user time to gain access to the control panel to unset the system. If the user enters through any other point including an access route then a full alarm will be generated. The system can be programmed to set after the exit time has expired or by operation of a final exit detection device.

Figure 5.10 shows eight detector circuits that also have their own associated tamper circuits. Each detector circuit and tamper loop is a closed circuit, but normally open devices can be connected across the CCT terminals and the A/T terminals. This layout shows the end station to which a number of keypads can also be connected in a star or daisy chain configuration to the three terminals +V, SIG and 0 V. Although the keypad only needs three wires connected to function with this control panel, additional terminals are provided in the keypad for two wire connection (0 V and I/P) for an exit terminate button or keyswitch, and for a lock release connection (+V and O/P).

The keypads themselves are programmed with a unique address using DIP switches located on the circuit board (Figure 5.11). The main circuit board also shows two programmable high-current transistorized outputs rated at about 200 mA each and identified as O/P1 and O/P2. These have either positive applied (+12 V when activated) or positive removed (+12 V removed when activated) states but with a number of different types that can be selected as listed in Table 5.5. The main circuit board, in addition, shows terminals for a total of two extension speakers (LS), which should not be placed in the end station.

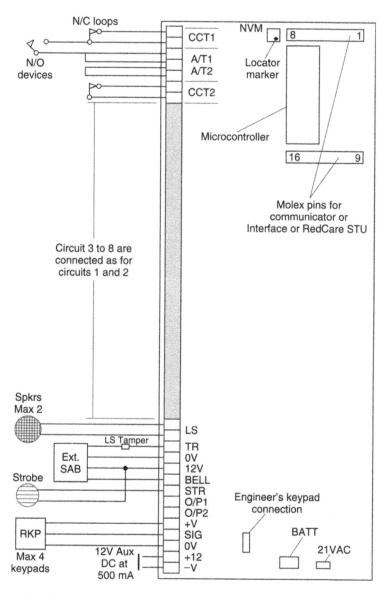

Figure 5.10 *Conventional terminal layout: separate tamper loop*

As an extension of the facilities offered by the first control panel that we considered, this later panel has Molex pins designed to accommodate a digital communicator, a RedCare transmitter or a specific interface board for use with direct line equipment or other types of communicator. The

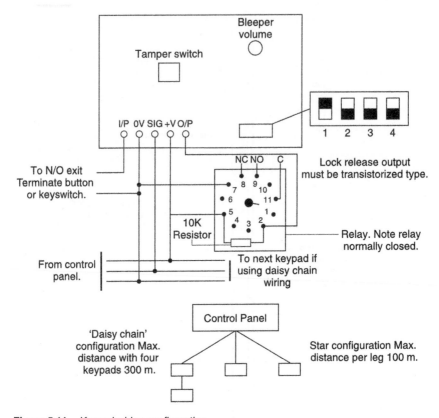

Figure 5.11   *Keypad wiring configuration*

interface board itself provides eight sets of changeover relay contracts for fire, PA, intruder, open/close (arm/disarm), trouble, low battery, technical alarm or alarm signalling.

The two control panel types that we have described are termed conventional, in that they have all of the detection circuits wired directly to the main circuit board. There are control panels that are conventional but can also be programmed to accept dual-operation wiring as described in Chapter 2 when the detectors employ end-of-line devices so that the alarm and tamper signal can be carried on the same two-core cable. These are not to be confused with multiplex techniques which use a data collection network.

The two control panels depicted in Figures 5.9 and 5.10 have detection latch facilities at the display, and so can be used to latch detectors that would otherwise switch momentarily and then reset causing their operation and alarm condition to be lost.

Table 5.5 *Output types*

Zone type	Description
Bell follow	This output will follow the state of the normal bell output, i.e. bell delay times and bell stop (duration) times. The output can be used for a second external sounder such as a high-power SCB (removal of supply on alarm) sounder, leaving the normal bell output for a standard SAB bell type
Exit/entry follow	This output type permits the addition of a separate entry/exit buzzer, or any other device, including lights, which requires to be activated during the exit/entry times
Set latch Type 1	This output type is active upon arming of the system, and remains active until the system is disarmed. It is also active for 0.5 second when the panel is reset
Set latch Type 2	This output type is active upon arming the system and deactivates when an alarm condition occurs or when the system is returned to the 'day' state. The output is also active for 0.5 second when the panel is reset. The output is active during engineer and user walk test
Shock sensor reset	This output type is designed to operate with all makes of stand-alone shock sensors. The positive power supply connection should be connected to this output. The output activates as a 6 second pulse at the beginning of the exit time
Walk test	This output is active during both the engineer and user walk test and the period between silencing the system and resetting the system. This output type is used on movement detectors which have the facility to switch off the walk test LED in any state other than a walk test
Entry system	This output type activates for 5 seconds following the recognition of a valid access code. The output will operate irrespective of the system status. This output type is designed to switch electric door releases. (Do not exceed 200 mA current drain)
24 hour alarm	This output is active every time a zone/circuit, which has been programmed as a 24 hour zone type, opens and deactivates when the zone is closed

It also follows that electromagnetic relays can also be employed to drive additional sounders or visual signalling devices by the use of extra power supplies with the coil energized from the trigger and hold-off terminals in the alarm condition.

The systems may be enhanced by the addition of certain ancillary control equipment.

5.4   Ancillary control equipment

There are certain items of equipment that we class as ancillary because they are not found in every system and can differ in construction depending on the manufacturer and control equipment type.

Remote keypads

For increased security the end station or main circuit board and power supply is best sited in a non-visible position in a secure area, with the facility for the system to be controlled from a network of remote keypads. These keypads can be fitted at convenient points close to the final exit positions. They can have a liquid crystal display (LCD) with the alphanumeric library and event log in English text or 'starburst' (Figure 5.12) or they may have seven-segment or individual LED displays.

   For aesthetic purposes they may have ultrasmall keypads, backlighting, hinged lids or flip covers and show the time and date. In certain fully addressable security control panels the system can comprise a control unit, remote input/output modules and remote keypads along a two-wire data communication line that is capable of measuring voltage and resistance in each of the zones as well as carrying out communication level diagnostics on keypads and on the subject modules. Remote arming stations are also classed alongside remote keypads, but these only allow setting and unsetting procedures to be followed and not programming or other system options.

Set/unset key fobs and tag readers

Keypads have always been a popular management and control setting device for all intruder alarm equipment.

   When mounting keypads it is worth remembering that they are best located with the display around 10 cm below average eye level. Fobs and tags bring along other considerations.

   For some time a variant system control method has been by use of the infrared key fob with encryption built into the key so that duplication was virtually impossible.

LCD English text   LCD starburst

Figure 5.12   *Remote keypad LCD displays*

Separate keys were available for the different users so that the system could be controlled either directly at the control panel or via small remote infrared receivers.

With the infrared key only basic alarm functions were possible as for full access to the options the keypad must be used.

As an extension on these developments control equipment has increasingly been found to provide an option on a solution for DD 243 set and unset requirements. These offer a remote keypad, internal and external readers and proximity fob. This all enables the flexibility to set and unset a system from outside of the premised according to DD 243 by means of a simple procedure. It will be found that this type of setting device is normally only compatible with equipment from a common manufacturer. They tend to be advertised as having tag proximity capability.

In order to overcome part guard options with key fobs they may be coloured differently on each side. For example red on one side and grey on the other. By showing the red side to the reader the system can be part set by the user when retiring at night. By presenting the grey side of the fob to the reader the building can be fully armed.

Using fobs with unique codes personnel can be programmed onto the larger system as an individual user, thereby providing complete identification of users setting and unsetting the system.

As integrated systems have come more on stream a greater use of tag readers has been introduced and popularized with tags having the same system LEDs as the keypads. Codes which could be four, five or six digit or tags could be programmed in a number of different ways such as:

USER
MANAGER
MASTER MANAGER
DURESS
GUARD CODE
ENGINEER

Additionally tags could be restricted to specific areas and to set or unset only. Therefore as a system schematic the tag reader would typically be configured as:

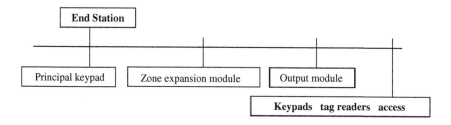

Keypoints

These are keyswitches that use the keypoint circuit to fully set or part set the system. They will normally be three position and have an LED indicator to indicate circuit faults when setting and part setting the system. The keypoint can be used with a time switch to set and unset the system; for instance, to switch on the alarm automatically at 18.00 hours and off at 8.00 hours from Monday to Friday and then to remain set at the weekend.

Printers

These are used to generate a real-time print-out of the system log. They are plugged directly into the control panel or connected via an adaptor and interface lead.

Permanent logging printers are intended to be sited alongside the main control panel and may be programmed to print a continuously updated record of events. They comprise a printer mechanism, a mains transformer with a fused mains supply input and a small power supply circuit board. They are mounted within an enclosure to match the control panel. Engineer-portable printers can also be found, and these are used to obtain records for engineering reasons. They may be used by the customer if the operator has a master access level.

Shunt locks

A further method of providing limited access to secure areas for the convenience of such persons as staff cleaners, who are not permitted to unset the full system, is by shunt locks. The lock must be sited adjacent to the entry/exit door where the buzzer can clearly be heard outside of the premises. These devices can also be used to leave one circuit set while the main alarm is switched off, by providing connections through a 24 hour module, in which case the indicating equipment must be sited in the main security area.

These units provide a good level of security yet are proof against false alarms caused by operator error.

Figure 5.13 shows a typical shunt for use with a specific control panel from the same manufacturer.

Considerations

- The entry door must be fitted with a key-operated switch.
- The shunt lock must only be capable of operation from outside of the controlled area.
- The door must have a switch contact.

Figure 5.13   *Shunt lock*

Operation

- *Entry*: unlock the entry door – this will automatically deactivate the system and permit entry.
- *Exit*: close and lock the entry door – this will activate the buzzer; press the setting push button – the buzzer will silence and the system set.

Pass switches

These are simple key-operated devices that are supplied in a small steel enclosure which houses the rotary switch. The enclosure is tamper protected. As the name implies, they are used to pass through an area protected by a detector. They do not have any LED indication and tend only to be used to isolate a door as a temporary measure when access is required through it. In practice, the pass switch switches across the alarm contacts on the detection device and includes the tamper loop. They should be used with discretion as they provide no indication except for the visual physical orientation of the key barrel mark.

Zone omit units

These allow either a monitored alarm zone to be isolated or two monitored alarm zones to be isolated (Figure 5.14). The omitted condition is indicated at the zone omit unit, but the isolated zone can only be switched back to the normal condition provided it is not in alarm.

Figure 5.14 *Zone omit unit*

These devices can have an internal piezoelectric sounder, to indicate isolate or fault, and terminals are provided for the connection of remote LEDs and extension sounders. They may be operated by a key or by a code. The wiring can be double-pole or end-of-line resistor.

The wiring is brought from the detectors to appropriate terminals in the zone omit unit, and these are then taken to the control panel. The zone omit unit is powered by the auxiliary supply of the control panel with the sounders and LEDs configured on unique terminals.

Zone splitters

These are used to indicate, at the unit, activation or fault during setting when two detectors are configured on one zone when it is the only practical method of wiring (Figure 5.15). The wiring method will depend on the control panel type, but it is connected as a direct interface between the detectors and the control panel zone.

Exit terminator

This may be a simple push button or a sealed prewired assembly that also contains a miniature speaker and an LED indicator. The use of exit

Figure 5.15   *Zone splitter*

terminators is the recommended positive false alarm free, setting method for panels with an exit terminator input.

5.5   Intelligent device (iD), multiplex, data bus networks and addressable systems

iD technology

First of all we know that we must carry out a risk assessment in order to determine a risk grade and then install equipment and component compliant products in relation to that risk. This has a great influence on the type of detection system that is to be employed. We also know that as we move through the grades starting with the lowest Grade 1 we encounter various wiring configurations – essentially single pole, double pole 4 wire, end of line EOL, dual operation fully supervised loop, triple resistors and multiplex data bus addressable configurations.

However, the question often asked is what is the role of iD systems in the European Standards and throughout the security systems industry in general?

We also need to ask how flexible iD systems are and by what methods are detection devices cabled in the industry?

In practice iD networks use a strict protocol parallel wiring bus which simplifies the system wiring and reduces cabling, but they are more prone to electrical interference than conventional systems.

iD is an active technology generally used as Grade 3 and is not passive in the same way as end-of-line or double-pole wiring. Using a parallel bus cabling format it simplifies and reduces the cabling. Easy system expansion is made by extending the wiring from any point. Cables are simply joined colour to colour.

However, it is fair to say that in situations such as take over sites it may not be suitable as it uses unique wiring. This will be specified by the manufacturer for the different cable purposes but it is normally screened. All manufacturers use the same iD wiring techniques.

When installing screened cabling it should only be terminated at one end of the run and this should be at the control panel end. The screen should be terminated immediately when it enters the box and only feature a short pigtail, as a long pigtail cancels out the effect of the screening. In those instances where the cable goes in and out of detectors it may be best to join the screens outside of the detector housings. It is preferable to terminate the screen outside of the box than to have a long pigtail inside of the housing (detector or panel). The pigtail is connected to metalwork using an appropriate point in the enclosure. The cables must not be looped inside the housings or routed under printed circuit boards. Throughout its length the screen must remain continuous.

It will also be appreciated that it is always more difficult to fault find in parallel wiring schemes. Equally active clocking bus technology which is an iD technique is inherently more prone to interference than conventional systems.

iD is not a detection system in itself but a means of simplifying wiring by individually addressing standard detectors that are cabled in parallel.

The iD lines are terminated using capacitors and silicon microchips (known as biscuits) are wired direct into the detector terminals. If it is not possible to wire direct to the detector terminals an iD junction box must be used.

Each biscuit has three wires coloured yellow, blue and white:

> **Yellow iD +**
> **Blue iD −**
> **White switch wire**

The yellow wire goes through the tamper switch and the white and blue connect to the alarm switch. Any form of parallel wiring can be used although twisted pair is not to be used.

It is possible to wire up to 30 biscuits on a single 4 core cable. The biscuit can report alarm and tamper over 2 cores and provide the power over the other 2 cores of the 4 core cable. This means that great savings can be made in cabling installation practices.

The iD points must not be duplicated.

It is the iD line that is used to check the status of the individual biscuits. This as mentioned is referred to as 'clocking' but is affected by cable resistance runs in excess of twenty ohms as this reduces the clocking signal.

iD is an addressable detection technology in which up to
30 'biscuits' may be located on a single 4 core cable
The iD line is constantly clocking the individual biscuits
to ascertain their status

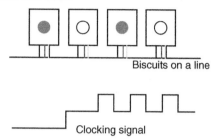

Biscuits on a line

Clocking signal

Figure 5.16 *iD biscuits on a line*

Each biscuit has a
unique iD number

Connections of biscuit
to detector and iD bus

• Yellow wire is iD +ve
• Blue wire is iD –ve
• White wire is the switch wire

• The yellow wire goes through the detector tamper switch
• The white and blue wires connect to the detector alarm switch
System can report tamper and alarm over 2 cores
The power is on the other 2 cores

Figure 5.17 *Connection of iD biscuits*

Therefore an iD system can be seen to continuously poll for two signals:

● Diagnostic to show devices are connected correctly are functional and tampers are closed
● A signal to report the status of the detection being monitored

Figures 5.16, 5.17 and 5.18 illustrate the standard industry detection wiring principles.

During the installation all the biscuits are to be installed inside the detector housings and the white wires are never to be lengthened to suit the application.

Always use opto-isolated iD junction boxes for multiple door contacts, foil, lace or roller shutter contacts as this prevents RF pick up in the biscuit white switch wire. See Figure 5.19 for installation logic.

The iD cable needs to be terminated with 0.01μf capacitors

The system specification must detail the end of each separate cable run or any spurs with more than two biscuits

Last biscuit

Yellow

White

Blue

Terminating capacitor

Figure 5.18   *iD cable termination*

All biscuits should be fitted inside the detector housings
The biscuit white wires are never to be extended

White

Door contact

Yellow   Blue

iD junction box
(opto-isolated)

Door contact

Figure 5.19   *iD installation of biscuits*

The commissioning involves measuring the resistance of the iD + and − connections at each detector to ensure it is correct for the cable length involved (typically 100 m of alarm cable is 16 ohms). Supply voltages at powered detectors should be measured and there is also a need to record the resistances of the iD + and − wires at the end station and compare these values with a specific equipment table to ensure that the correct number of biscuits is being read. Readings should also be taken to the mains earth to check for earth leakage on both iD wires and these wires should also be checked for 0 V/12 V shorts. Although the commissioning is rather different to that which must be carried out with conventional systems it is still understandable. However, a further reading must be verified with certain control equipment as it is a critical voltage. As we know the iD + is the clocking signal that addresses the biscuits. On the iD bus there are a number of signal levels and in some cases a mid clock voltage value. This must meet a quoted value or be adjusted to suit or the biscuits will be miss-clocked and not report correctly.

Fault messages and indications are given but additional resistors can be used to find faults such as open/closed/tamper/detector fault/detector mask. iD line shorts may be shorted to itself or to a supply connection or to earth.

Surveying Guidelines

In view that iD is an active technology with data being clocked up and down the data bus there are a number of guidelines to follow. Failure to do so will result in false alarms and remedial work.

It becomes clear that in the first instance a survey must be carried out to establish if an iD system is a good proposition for the premises to be protected.

If it is not an option to use iD fully throughout an installation it may be possible to use a mix of technologies using end-of-line wiring (EOL) together with other devices using iD. Control equipment is available from numerous manufacturers to give the installer total flexibility in this regard. iD points and EOL zones are designed for the distribution of zone inputs and outputs and can be used side by side in most units of control equipment.

Due to the specific nature of iD techniques there is merit in recording the most important aspects which can act as surveying guidelines. These are as follows:

- Screened cable is always to be used. This applies even to a greater extent in 'noisy' environments
- Using 4 core cable no cable run should ever exceed 100 metres
- Cable resistance in excess of 20 ohms will reduce the clocking signal
- 6 core cable with the iD cores doubled up will permit runs of 150 metres. Doubling up supply cores is beneficial to minimize volt drop but is not to be used for communication connections.
- Never run the cables alongside cables carrying AC or digital signals
- Never use any other single cores in the same cable such as a speaker, keypad or an iD line from another panel or a zone expansion module.
- Always use opto-isolated iD junction boxes for multiple door contacts, foil, lace or roller shutter contacts as this prevents RF pick up in the biscuit white switch wire.
- iD biscuits are sensitive to strong magnetic fields and high energy RF radiation so specify an opto junction box outside of the area to protect the biscuit.

- The strategic positioning of zone expansion modules can reduce the problems associated with long cable runs. In the larger building application specify the actual location of the zone expansion modules. These can be 1000 metres from the panel on the RS485 data bus. The zone expansion modules should not be clustered around the panel but sited to collect zones locally around the building.
- Zone expansion modules that have their own power supply should all be configured on the same electrical phase in buildings with multi-phase systems. They should also be on the same mains phase as the end station. This is to be detailed for the installation engineer.
- Extreme caution must be taken when considering the possibility of using an iD based system to replace a conventional system. This is because the wiring may be old and in a poor condition, making it unsuitable for modern addressable signals. A thorough site survey must be carried out.
- Specify opto-isolated junction boxes in instances such as: Where a number of remote zones return to one junction box; the wiring distance exceeds 100 metres; the wiring is in poor condition or where it runs through areas with machinery. In some cases it may be advisable to use a mix of zones with some iD and others EOL or specify a system with end-of-line devices only.
- In terms of the cabling there is a need to avoid high resistance in junction boxes and in detectors. Ensure that there is no earth leakage between the iD line and earth and no voltage shorts onto the iD line.
- Fit a terminating 0.01 micro farad capacitor at the end of each cable run across the yellow and blue wires and spurs with more than two biscuits.
- Filtering options are available with many units of equipment to minimize the chances of false alarms. They work by increasing the number of scans for tampers and twin devices by a given numerical factor.
- It is possible that tampers (missing biscuits) and twin devices (two biscuits reporting in at the same address) may occur. This can be as a result of interference causing a biscuit to count its address incorrectly or because of a long cable run.

Multiplex offers high levels of security and reliability while reducing installation and maintenance costs. With conventional wiring systems it was seen that each detector circuit runs on a pair of wires for alarm detection; however, the multiplex method uses the main data highways or branches from the main control panel. This means that a huge reduction in cable installation is achieved. Using a multiplex technique each branch can comprise, as a minimum, four-core cable which can be of the order of 2 km in length and can typically accommodate up to 16 line interface modules (LIMs). These LIMs can have five double-pole circuits which may be individually programmed to a circuit type with appropriate attributes. These parameters will not be unlike their conventional wiring counterparts.

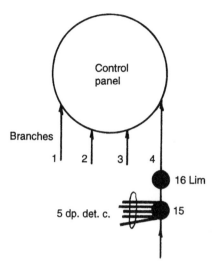

Figure 5.20 *Multiplex technique*

Consider as an example a system with four detection branches which will have a capacity for 320 double-pole circuits all individually reported and monitored (Figure 5.20). Figure 5.21 shows the main circuit board connections with the wiring to the LIMs and detectors including the individual tamper loops (A/T).

The student will note that branch 2 has a six-core cable, as the detectors in this LIM must be powered or a remote keypad may need to be sited. The LIMs themselves may be boxed and complete and with a printed circuit board cover and box tamper protection or be supplied bare for fitting within a power supply unit with a tamper protection facility. These power supplies are used to boost the branch line voltage when it drops below 12 V.

The engineer will note that in this example of multiplexing that the detection circuits do not have any end-of-line devices. The next stage is to look at a variant data bus network addressable system which employs a slightly different approach and is referred to as expandable (Figure 5.22).

This system has eight zones on the main control board and two zones on board each remote keypad. These zones may be closed loop or employ end-of-line resistors. Each remote keypad also has a programmable output. If further outputs are required, then local expansion cards (LECs) may be fitted to the remote keypad network to provide two additional zones and one output. The control unit runs two separate data bus networks, one for the remote keypads and the local expansion cards, and a further data bus network for nodes. These nodes can number up to five on a network, and they themselves provide eight zones and two outputs. An optional iD, ID or I/D referenced node allows ID-compatible detectors to be used with the

Figure 5.21   *Main multiplex circuit board*

panel. This ID node provides a single ID detector loop for connection to 30 ID devices and eight programmable outputs.

This system can be seen to be expandable from 10 to 56 zones. The alternative ID node allows an alternative method of configuring the zones and can accommodate any normally closed detection device in conjunction with a discrete ID sensor. The ID sensors are wired in parallel across a sensing line which may be of T, star or ring format. The cable is two core for sensing only or four core for sensing and voltage supply. ID technology is a method of sensing and transmitting signals using an advanced silicon microchip at addressable detector points. It provides individual identification of every detector on the network using only a four-core cable to also cover anti-tamper. These ID detectors can be used on a range of ID-compatible control panels, which can analyse the signals from any device due to the parallel wiring configuration.

At present, ID sensors are less widely used than the traditional detectors, but they hold great potential for the future. In any event, huge flexibility exists in the standard node wiring method (Figure 5.23).

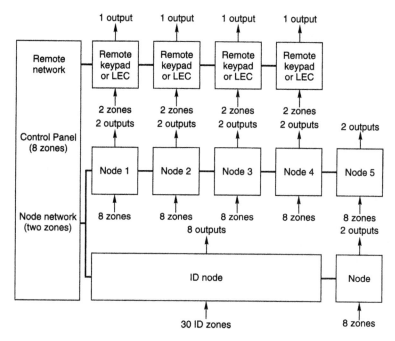

Figure 5.22   *Expandable system with a node network*

Figure 5.23   *(a) Node system layout; (b) node terminal detail*

The cable can be a daisy chain and/or a star pattern using four-core cable with detectors being connected as in Figure 5.24 using end-of-line or double-pole wiring. This applies equally to control panel, remote keypad or LEC terminals.

Observations on EOL resistor wiring are:

- the detector alarm contacts must have a 4k7 Ω shunt resistor fitted;
- a 2k2 Ω end-of-line resistor must be installed at the furthest point;
- the loop resistance with the end-of-line resistor shorted must be less than 100 Ω ;
- the maximum number of detectors allowed on one circuit is 10.

Normal British Standard conventions apply to the circuits. Therefore, up to 10 door contacts can be connected to each circuit, but only one movement detector such as a PIR.

With the different levels of equipment and different means of gathering information from the network, the identification of the circuit must also

Figure 5.24   *Node detector wiring*

surely change. Consider an example using LIMs. Each circuit is allocated a unique four-digit identification number, LIM number and circuit number as follows:

1st digit	Branch number (1, 2, 3 or 4)
2nd and 3rd digits	LIM number (01 to 16)
4th digit	Circuit number (1 to 5)

Example. Circuit 3 on the fourth LIM of branch 2 is identified as 2043.

LIMs must be numbered in sequence for each branch, i.e. the first LIM on each branch is numbered 01. On the LIM circuit board the in and out connections must be followed to achieve this.

When using nodes, the node network cables must be connected to the appropriate terminals and the detection circuits connected to the appropriately lettered circuit (Figure 5.24). The node I/D selector switch must then be set. This should be set sequentially to aid fault finding and programming. The circuit can then be established by reference to a table such as Table 5.6.

After initial power up or in an unset condition, the intruder alarm engineer can scroll through the engineers menu by entering the engineer's code. The circuit types and attributes can then be programmed to suit the protected premises.

In the field a vast range of equipment is available to satisfy different multiplex techniques, and the modules employed will be allocated different terms and defined in various ways. Examples are multiplexed input and output modules (MIOMs) and remote input/output (RIO) modules. The method of data collection and capacities will vary, but the concept of using data highways extends throughout. However, it is important to pay attention to the manufacturers' data in respect to the cable to be used since RS485 wiring for keypad connections etc. or ID wiring may invoke a specific type and it may not be possible to use standard alarm cabling.

Table 5.6   *Node identification circuit*

Node I/D	Detection circuits							
	A	B	C	D	E	F	G	H
1	17	18	19	20	21	22	23	24
2	25	26	27	28	29	30	31	32
3	33	34	35	36	37	38	39	40
4	41	42	43	44	45	46	47	48
5	49	50	51	52	53	54	55	56

5.6   Voltage surges and induced electromagnetic energy

Voltage surges

Voltage surge is more common than we may think, and has many causes, and today the price of damaged equipment and lost data is greater than ever. These microsecond-long surges in nominal system voltages have caused difficulties for manufacturers, installers and users of electronic and electrical equipment for many years. Voltage surges cause serious problems by corrupting or erasing data and damaging equipment, and are responsible for a huge loss of revenue as a result of down time. In certain cases these transients can be life-threatening.

In practice, voltage surges occur millions of times a day on typical AC power and communication lines world-wide; therefore, it is vital that measures are taken to prevent damage as a result of them. At the present time this is particularly apparent, with increasing numbers of companies depending on the potentially highly vulnerable circuits of computers and security systems.

Uninterruptible power supplies (UPSs) will not protect sensitive equipment against all transient surges, and these themselves may need protection.

Typical causes of these voltage surges range from lightning and electrostatic discharge to inductive load and power grid switching. Equally, many everyday objects such as lifts, photocopiers, fluorescent lights, vending machines and computer printers can cause transients. The damage that is caused to unprotected electronic equipment varies from slow degradation in performance to the sudden destruction of an entire system. All security systems are potential victims, and even in the domestic sector with the advent of telecommuting and the increased use of personal computers (PCs) and modems connected to security control equipment the risk of damage from voltage surges is becoming more real, as are the associated costs should a problem occur. To this end, protection should be applied to all sensitive electronic equipment connected to the mains power supply and modem lines entering the building.

BS 6651: 1992, *Protection of Structures Against Lightning*, addresses the problem in Appendix C. This increases the scope of the standard to include guidance on the protection of the equipment in which we are interested, and gives advice on assessing the level of risk. It also recommends methods of protection and the selection of protection devices. This appendix also takes account of the effects a lightning strike can have for buildings with external power and data feeds and how they may be affected by strikes anywhere near these cables. The appendix is concerned with cable runs up to 1000 m from the building.

Table 5.7 *Magnitudes of surge voltages and currents according to BS 6651*

	Category		
	C	B	A
Peak voltage 1.2/50 µs waveform	20 kV	6 kV	6 kV
Peak current 8/20 µs waveform	10 kA	3 kA	500 A

There are effectively three areas in a building with different protection requirements:

- The supply side of the incoming power distribution board — Category C
- The mains distribution system — Category B
- The load sides of socket outlets — Category A

It should be noted that category A may not be applicable in certain buildings since the socket outlets may be too close to the distribution board, and in such cases a category B protector should be used.

The magnitudes of surge voltages and currents defined by BS 6651 that should be protected against at different locations on an arc power system and in an area of high system exposure are given in Table 5.7.

System exposure is affected by several factors such as geographical positioning and the level of loss inflicted if a lightning strike was to cause damage.

It is a requirement of BS 6651 that all telecommunication and data communication equipment must withstand those surges defined in category C, i.e. 10 kA peak current for high-risk system exposure, but this is reduced to 5 kA for medium-risk system exposure areas. The reason for this is that the electrical characteristics for power lines and signal wiring are different in that a surge travelling in a signal cable is not attenuated by the cabling in the same way as a power line surge. Clearly it is the surge protector which must withstand these peaks to ensure that the equipment continues to operate, but it must itself survive the transients and resulting currents, allowing through only the residual part of the surge which is not capable of causing any damage.

The residual part of the surge is called the 'let-through voltage'. Most electronic equipment will withstand short-duration surges of around 1250 V on the mains supply, and a reasonable safety margin is a let-through voltage at the equipment interface of not more than 900 V. Clearly the closer the let-through voltage is to the equipment supply voltage then the better equipment protection.

The Loss Prevention Council (LPC) is used by insurers to assess risks and avoidable claims in the event of damage caused by surges and spikes. This highlights the importance that is being placed on the subject and the growing need for suppression of the problem.

Mains protection systems

There is now a wide range of transient voltage surge suppression products available, ranging from surge protection systems based on a trilevel zoned protection strategy that starts where power and communication lines enter a building to special protection devices for the most sensitive equipment. The heart of these systems is a distribution surge protector (DSP), which is installed at the cable entry to a building and at the distribution point for each floor of a building. These may be in single- or three-phase guise. There is also a range of devices available that can incorporate power and modem protection in one unit.

In Section 9.3 we look at false alarms and mains-borne interference and the need to use filters in the form of fused spurs with transient voltage suppressors. These are intended to prevent this distortion of voltage, current and frequency as a result of transients. These devices should have certain properties:

- high energy resistance;
- high-speed suppression;
- solid state, i.e. no moving parts;
- isolation of the primary from the secondary side.

It will be found that there exists a range of filtered mains spur units available for the intruder industry, and if these are used local to the control equipment together with a transient surge protector wired across the building mains supply, the problem of mains interference and its resultant damage should be negated.

Other practices that can be followed to reduce surges and thus the incidence of false alarms as a result of environmental interference can be included in any system:

- Adding surge arrestors on the secondary sides of transformers. This can also apply to the sensor wiring to inhibit any excess voltage or current.
- Avoiding any external wiring.
- Physically connecting all steel conduit tubing to a good system ground point.
- Providing a separate ground heavy gauge insulated earth to the main transformer of the system.

Data line protection

The intruder alarm engineer is having to become increasingly involved with remote signalling and data transmission, and should be aware that the protection of data calls for a somewhat different approach to the protection of power lines, but is just as critical.

Standard hybrid surge protection devices are usually not designed for high-bandwidth data transmission as well as significant surge diversion capability. Also, the need for low clamping voltages and delicate impedance matching makes circuit design particularly difficult. This applies even more when high transmission speeds are specified.

Specialized protectors are, however, produced for telecommunication areas and modems, which can be severely damaged or even destroyed by large transient surges. To this end, unique protectors are available at telephone outlet sockets intended for modems. The use of other protection devices is advocated where data is to be transmitted via networks between buildings.

In summary, with the increased use of complex electronics and remote signalling plus the introduction of diverse voltages and frequencies into buildings, the protection of equipment has become more important. In addition, however, to transients on the mains supply there are two other major types of interference that can compound problems, namely RFI and EMI.

Radiofrequency interference (RFI) and electromagnetic interference (EMI)

RFI is interference induced by sources of radiofrequencies such as transmitters, CB radios, or discharge lighting, whereas EMI is generated by radiated electromagnetic energy as mains-borne interference. EMI sources include welding tools, the starting/stopping of motors and lightning. Although all electrical equipment is capable of causing a spike when switched, EMI is often generated by fluorescent lights, air conditioning units and, particularly, equipment containing motors.

Both RFI and EMI distort the current flow, particularly DC which in the case of a microprocessor is seen as a pulse of instruction which results in a system crash or memory erasure.

Protection against interference is very much the responsibility of manufacturers of electronic equipment, who incorporate components to suppress the problem. However, measures can also be taken by the installer, and include:

- Primary filtering of the control equipment.
- Screening and shielding by the use of grounded conduit tubes or shielded cable. Additional bonding and earthing so that equipment is shielded by metalwork, and the impinging radiated energy is conducted to an earthed point.
- Secondary filtering of the detector and signal wiring.
- Segregation from other cabling in a building.

In practice the installer can consider the use of filters designed to allow only the transmission of recognized signals within the system with unwanted signals being suppressed. This involves the application of filters on the primary side of the control panel and also on the outgoing cables.

Screening and shielding is essentially the enclosure of system parts in conduit and metallic structures which will ground radiated energy travelling through the atmosphere. Screening is achieved by the use of braid in cable, seen as a screen of fine wires which is grounded to extra earth points.

Segregation is simply the use of exclusive routes for alarm cabling to ensure that harmonic currents from other cables, particularly power wiring, cannot be induced in the alarm cable. Alarm cable should never be run close to power cables or in parallel and should only ever cross at right angles, or else must be additionally shielded by an appropriate earthed metal division.

In summary, we should look for third-party verification that equipment has resistance to interference and will be tolerant of radiated RFI and EMI to a level acceptable to the application in hand. Also, steps to minimize the effect of RFI and EMI should be taken:

- ensure that the control panel and all connecting wiring is screened from the effects of radiated RFI and EMI;
- check that the system is properly fused and grounded;
- ensure that the system is filtered as appropriate;
- make sure that the system has been installed with consideration given to lightning strikes.

Static electricity

A final consideration is static electricity, which can damage sensitive electronic components if it is discharged into them because a control panel has not been correctly grounded. Static electricity, if allowed to generate in a panel, can easily cause random faults and alarm conditions that can be extremely difficult to identify.

Available equipment

12 V spike suppressor

This is intended to eliminate induced AC voltage and prevent false alarms caused by spikes and lightning.

An initial test involves taking a reading across 12 V DC+ to the energized detectors and the mains earth to ensure that the value is not above 1.2 V AC.

If a high induced AC voltage is measured, a suppressor should be fitted to the control panel and any power supply (Figure 5.25). Once fitted, the value decreases to about 0.5 V AC.

Figure 5.25   *Spike suppressor*

Installed as shown, suppression will eliminate induced AC and electrical spikes picked up by alarm cables acting as aerials to detectors, LIMs, keypads and such. In practice, no matter how far alarm cables are segregated from mains wiring, induced electrical signals from fluorescent lights, fridges, freezers, central heating pumps and motors can affect the 12 V supply to control equipment and alarm detectors. This will lead to false alarms, system malfunctions and communicator activations.

The suppression is attained by the decoupling of induced AC from every alarm cable connected to the panel and stabilize the 12 V DC supply to spikes which would otherwise affect the control equipment, alarm detectors and communication system.

PIR stabilizer

This is specifically designed to prevent unexplained PIR false alarms caused by EMI/RFI signals induced in the alarm cable acting as an aerial (Figure 5.26). These stabilizers incorporate a capacitor to stabilize the 12 V DC regulated supply against ripple and to power the PIR for some 500 ms. A varistor protects the DC supply from transient spikes, and an inductor blocks the passage of RFI from the cable into the detector, with a diode being employed to ensure maximum stabilizer efficiency.

The installer can carry out normal checks for a troublesome PIR:

- Is the DC voltage stable and of the order of 13 V DC?
- Is there any infrared energy movement in the field of view?
- Is there any direct draught or are cable entry holes unsealed?
- Does the detector feature a white light filter to negate reflected light?
- Has the PIR been securely fitted to a stable surface?

Having carried out the above checks, if the problem still remains, the cause may be an EMI/RFI signal.

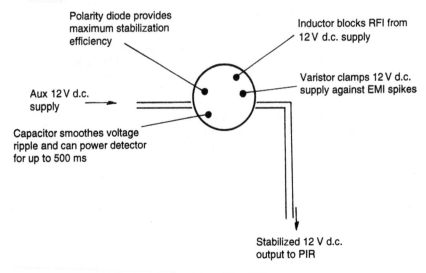

Polarity diode provides
maximum stabilization
efficiency

Inductor blocks RFI from
12 V d.c. supply

Varistor clamps 12 V d.c.
supply against EMI spikes

Aux 12 V d.c.
supply

Capacitor smoothes voltage
ripple and can power detector
for up to 500 ms

Stabilized 12 V d.c.
output to PIR

Figure 5.26   *Stabilizer suppressor*

14–21 V a.c. to
control panel

N

Earth

L

Fused

Figure 5.27   *High-frequency earth line choke*

High-frequency earth line choke

This is intended for use with steel-boxed equipment to tie the 0 V line to earth as a preventative measure against RFI (Figure 5.27). Being connected with the incoming earth, the choke protects against high-frequency RFI signals leading back from the mains earth onto the 0 V line, which in turn affects the regulated supply and communicator and alarm detector wiring.

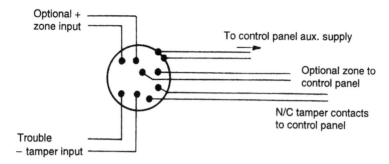

Figure 5.28 *Trouble tamper relay interface*

Trouble tamper relay interface

This is intended to eliminate tamper circuit faults caused by EMI/RFI affecting the 0 V tamper circuits (Figure 5.28). It can be used on bell tamper circuits as well as on standard or global tamper circuits. Troublesome closed-circuit tamper lines should be connected to the negative input of the interface, and any problematic N/C zone can be wired to the optional positive input if so desired. The device is then connected to the relevant panel circuit and regulated 12 V DC supply.

Four- and six-wire filters

Specifically designed to protect alarm data cables, these filters attenuate RFI causing signal corruption on panel keypad, LIM and speaker cables acting as aerials. Although they offer little resistance to conventional DC circuits they do substantially attenuate RFI in the 25–800 mHz spectrum, making them compatible with both analogue and digital microprocessor control equipment.

This type of filter can be combined to screen any number of alarm circuit wires, and will be found to offer far greater attenuation than screened cable alone.

These filters should be installed at the problem end of the cable such as the panel; to obtain maximum attenuation they can be fitted at both ends of the cable. It is important to filter all circuit wires (even unused spares) and also to include the 12 V DC supply.

The wires to be filtered should be disconnected from the control equipment and then connected to the filter terminals. The filtered output cable should then be connected to the control panel terminals. If required, this can be repeated using a further filter at the keypad, LIM or speaker.

The attenuation on using a filter is shown in Figure 5.29, using broadcasting, aeronautical use, military, police/emergency services, UHF, television and cellphones as typical examples.

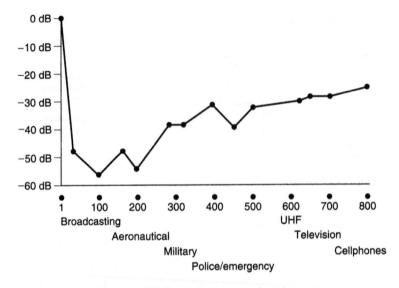

Figure 5.29   *Attenuation by an RFI filter (values in megahertz)*

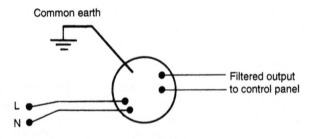

Figure 5.30   *Mains filter*

Mains surge purge filter

This will protect the control panel against both EMI and RFI via the mains supply when sharing a common supply with fridges, freezers, central heating pumps and motors or other highly inductive loads. It may be fitted inside the control panel or form part of an exclusive spur format (Figure 5.30).

Fused mains clamp suppressor

As the mains surge purge filter this can be fitted within the control equipment or be part of a spur unit. It will give protection against electrical spikes and lightning via the mains supply.

The internal safety fuse of the clamp suppressor should be replaced if it is blown by lightning.

With some knowledge of voltage surges and EMI and their prevention, it is essential that one overriding point be firmly made: under no circumstances should an earth route be modified to reduce inducted or radiated interference problems if it compromises the earth safety function. If this is the case, the installation should be reconsidered so that *both* safety and EMC (Electro Magnetic Compatibility) Directive requirements are fulfilled.

We can summarize by saying that an optimum configuration for complete electromagnetic compatibility of a known system is a completely sealed enclosure that acts as a screen to radiated interference, either in or out. In the real world this cannot be achieved because for the intruder engineer there are connections that must pass through the sealed enclosure. Also, the wiring connections can cause problems if good wiring practices are not observed.

The connections to and from an enclosure might be the necessary input and output cables of a system control panel, and an easy way to improve the electromagnetic compatibility of these entries would be to use screened cable.

The cause of induced current problems should be investigated with a view to prevention rather than cure. In some cases it will be better to rethink the installation of the alarm cabling to avoid the interference.

5.7　Networking. Uploading and downloading

Networking

Networking and information technology certainly do, and will increasingly include intruder alarm system functions. For this reason it is important that the security systems engineer does have some knowledge of the basics of networking.

For applications in the electronic security systems industry we should appreciate the following main points:

- A network in its strictest sense is a group of three or more devices that can communicate with each other.
- A data network is an interconnection of cables and equipment that carries data.
- Data is regarded as electrical signals that convey information.
- Routers are internet working devices used to direct packets (a group of bits) from one network to another.

Networking came along very much as a result of the need to share information. As the telephone and telecommunications sector has expanded so has the use of networking. The interconnections between the networks are of different sizes so do have separate identities.

A big network is known as a wide area network (WAN)

This covers large areas and because of the extended distances involved it often uses radio, telephone lines or satellite links rather than the traditional cable we are familiar with. They may also be connected to many medium sized networks which may at times be referred to as a metropolitan area network (MAN).

A small network is known as a local area network (LAN)

To qualify as a LAN it only needs a minimum of three addressable points or nodes. It is possible to connect LANs to MANs and WANs.

A prominent LAN technology is Ethernet. It allows the use of twisted copper and co-axial cables in various guises to be used to support the various data speeds. It has been adopted as an international standard by the International Standards Organisation (ISO) and can be easily expanded. Developed as an essential format for networking, it is often found in offices connecting together computers on a LAN. In view that it includes a wide range of data speeds, a naming convention has been developed to identify details of the system to illustrate the data signalling rate, transmission method used and the maximum segment length of the network or type of media used.

Originally Ethernet was recognized as a bus topology but tends now to be classed as a logical bus.

In order to make the connection between the control network and supervisory PC serial communication ports are used.

Serial ports on PCs and workstations in the main use the RS232 standard.

This defines the electrical and mechanical characteristics of the connection but successful communication can only be achieved by the software of the devices which are being connected to each other.

RS485 is a further standard for serial connections often used to connect controllers on a data bus.

Ethernet can be seen as a system that allows connection to be made to PCs and workstations by RJ45 ports.

Intruder alarms are being regularly developed in a matrix form so as to be built up in a network. This enables contact ID reporting via RS485 network. By this the control panel can still retain its independence and ability to communicate via built-in digital communicators and such.

A network type is shown in Figure 5.31. This system is typical of a network and has full up/download capability with true partitions and account codes per partition. It allows events, event history, record history, schedules, test calls and other relevant information to be displayed for all

Figure 5.31 *Network type upload/download and partitions*

system activities and this information can be accessed at the numerous points on the network. It is to be noted that systems can be subdivided for operating convenience and this is known as partitioning the system. Additional partitions may be set up as subdivisions of areas (or levels) by the use of *wards* or *shunts*.

Systems that send an alarm signal to private security central stations are now well known within the industry, and are used for most types of risk. The growth of these facilities is a testimony to the advantages of remote signalling. This transmission certainly optimizes police resources and allows operators at the receiving centre to make valued judgements and interrogate systems.

In the central station itself the software can be viewed as an expert system for handling data and then determining the responses that are required. Instructions can then be issued to people, with correct information being available at an early stage.

Software programs must of course be fast, simple to operate, and have inherent flexibility. There are many features that the software must contain at a minimum, and in the future this should lend itself to both uploading and downloading protocols.

Up- and downloading is the process of allowing a security system to be remotely programmed or interrogated via the telephone line and a PC. The ability to program security systems from a remote location rather than from the installation itself, offers tremendous advantages to the end user and installer. Clearly, although the central station will have the necessary programming capacity to perform this function, even the less complex and mid-capacity PC can perform the task via connection to a modem.

For the end user the downloading protocol offers, above all, peace of mind. No longer is it necessary to have a stranger on the customer's premises, even if the visitor is there to help protect the premises. Equally, the knowledge that assistance, in the event of something going wrong, is only a phone call away is reassuring for the end user. Indeed, if the requirements change the system can be easily reprogrammed at any stage to allow for additional keyholders to be added, change of use of any area, extension of entry/exit times, etc.

We can therefore define uploading and downloading in general terms as enhancing communication. A plug-in or boxed digital communicator will allow alarm status information to be transferred to a dedicated central station. With uploading and downloading techniques a digi-modem is connected to the control panel either as a plug-in or stand-alone device, and allows remote interrogation to be done via the PC. The digi-modem is also able to function as a standard digital communicator if so desired. The non-volatile memory (NVM) within the digi-modem can generally be programmed via the control panel, as can the modem data.

Figure 5.32   *Remote programming*

In Figure 5.32 we can see how a digi-modem in conjunction with a PC and compatible modem is configured to achieve upload and download by programming on a remote basis. In this case, as an alternative, the software can be installed onto a portable PC with a PC interface (PCI), allowing the data to be uploaded or downloaded from the control panel on site. The PCI is additionally supplied with a D-type socket which can plug directly into most portables. A convertor may be necessary if the PCI is to be connected to a desktop PC. An alternative option, where a portable PC is not available, is to use a dump box. By plugging a dump box into the PC the system profile can then be uploaded or downloaded. This information can then be downloaded to a control panel, and the system configuration will be automatically programmed.

It follows that in practice the engineer must first obtain confirmation of the compatibility between software packages and the control and signalling equipment. In general it will be found that a plug-in digi-modem must be used with a control panel from the same manufacturer.

The first system we will consider has the software installed on a laptop or desktop PC, depending upon the requirement. We can consider its principal features as:

- remote configuration and maintenance;
- integral customer database;
- optional on-site use;
- remote log dumps;
- on-line keypad mode of operation.

As an extension of this concept we can then consider how a number of independent software modules can be adopted and monitored at a PC-based monitoring and control centre (Figure 5.33).

In this case, individual systems can be customized to exact requirements. It is an effective solution for various security management applications, including control of sites with multiple buildings. Management and

Figure 5.33 *Monitoring of independent software modules*

monitoring from a head office of the branch security systems in banks and other financial institutions can also be effected.

The operation is controlled by the use of detailed graphical representations, including detailed map references, security configurations, locations within buildings, etc., based on Microsoft Windows technology with built-in help facilities. Customer histories are maintained on a system hard disk with encryption communications supported over the PSTN, Radio, redcare, Private Data Networks or ISDN.

The programming will include sequential alarm reporting and Point ID signalling to assist in alarm verification in compliance with ACPO directives.

The idea of partitioning systems is well known. Residential systems can be implemented with part set functions allowing only the arming of selected parts of the system by dividing the installation into, for instance, two parts. This way the perimeter protection can be armed when at home and a full set can be performed on leaving the premises.

For commercial and industrial applications the system can be divided into blocks; for example, one block for the warehouse and another for the office area. Both blocks can be armed and disarmed separately. A common area such as an entrance to a reception will only arm if both blocks are armed. It will automatically disarm if one of the blocks is disarmed. In order to only use one control panel or item of control equipment in applications which normally require more than one control panel, the system can be divided into splits. Each split will have its own dedicated keypad(s), zone inputs, communicator outputs, sounder outputs, etc. This allows the user to operate the system as if it were two separate systems. Control panels are available which can be divided into two splits, and others can be divided into four splits. Each individual split can be further divided into two blocks and a common area.

This type of system configuration makes the control panel very flexible, allowing it to be used for a wide variety of applications. These uses can range from simple and complex residential applications to high-security bank installations.

Various programming options and the possibility to expand these systems in terms of zone capacity makes them further suitable for large and complex industrial applications such as office blocks, hotels, government buildings, warehouses and shopping centres. High-security applications are possible by adopting true end-of-line resistor networks and extensive alarm reporting via dedicated digi-modem diallers. These then become an ideal solution for banks and museums.

MS-DOS-based monitoring software enables the monitoring of various sites, e.g. the main office of a bank can monitor all of its local branches. Large office blocks can integrate their control panels into a building management system. The software will allow remote operation, monitoring and alarm verification. Central stations can upload the alarm logs, arm the system, control the outputs, etc. Indeed, the MS-DOS-based program can also display maps of the various sites monitored and obtain an inside view of what is actually happening.

Specially designed digi-modem diallers that plug into specific control panels with a particular serial communications capability are available. These will often be of 16 channel capacity when operating in fast format and reporting to a digital receiver. In slow format this will be reduced to 8 channels. Networker systems, which are Microsoft Windows-based programs, offer real-time monitoring of a number of control panels on a single PC. This gives an enormous zone control capacity. Additionally, the flexibility of the Windows environment allows the user to simultaneously use other applications while maintaining full communications with all control panels. In the event of an alarm activation the networker system window will automatically 'pop up'. These networker systems support the Windows DEE protocol, allowing data to be exchanged from one Windows application to another. Therefore, these networker systems can be incorporated into other applications such as building and security management systems.

Each control panel can have one main map with additional submaps that are represented by relevant icons displayed on the computer screen. This introduces the added security of various guard tours, a unique feature providing safe and secure guard patrols.

The block system is divided into two blocks, one warehouse block and one office block. Both blocks share the reception as a common area. This common area will automatically arm when both blocks become armed. When the system is armed the common area will disarm when either block is disarmed. The LCD keypad will indicate that a part of the system is still armed.

Outputs

Keypads 1–16

Central station

Block system

Block 1

Common

Control panel

Warehouse area

Office area

Reception

Split 2 outputs

Keypads 9–16

Central station B

Split 1 outputs

Keypads 1–8

Central station A

Block and split system

Block 1

Common

Control panel

Warehouse area

Office area

Reception

Tenant

Split 1

This system is divided into 2 splits. Each split can have its own dedicated communicator outputs, sirens and keypads. Therefore a tenant can define its users, who can operate only the functions related to the tenant area. Split number 1 is divided into blocks and a common area. This could also be done for split number 2.

Figure 5.34 *Block system*

As an exercise, we should consider the block and split method and how uploading and downloading can be applied to cover future eventualities and the remote setting that would need to be employed (Figure 5.34).

Dedicated line. When PC dials panel response is immediate
Shared line. When PC dials panel it hangs up after second ring. This primes the panel
to answer the second call

Figure 5.35 *Architecture connection of PC to CIE*

Company Perspective. Benefits	
Alarm Company	**Customer**
Programming the system and any adjustments via the PC	Immediate response to requirements for changes
System problems reported to PC as they occur	Regular system verification
Remote interrogation of the system including diagnostics prior to reset after an alarm	Status interrogation for multiple sites
Remote servicing of the system and interrogation in advance of on-site servicing	Analysis of user irregularities
Remote access to logs to assist in the investigation of alarms etc. Savings in man hours on site	Storage of logs on PC. It is not normally possible to download data with the system set although logs and diagnostics can generally be uploaded
Functions available at the system keypads become available at the PC	Selected events are signalled to the PC

Figure 5.36 *Remote interrogation. Company perspective*

As a conclusion to this section it is possible to show the architecture of how the control and indicating equipment (CIE) is configured in a practical way by two different optional means in order to perform uploading and downloading. This is illustrated in Figure 5.35 with details of the different line modes. Remember the PC we are referring to is that which is accessible to the security installer.

Figure 5.36 shows how the benefits of remote interrogation via uploading and downloading from a company perspective apply to both the installation company and the system user.

6 Signalling systems and confirmed alarms

This chapter provides the reader with an understanding of the three different types of signalling available, namely audible, visual or remote. It also helps the reader appreciate the alarm confirmation techniques that are currently in force.

In the first instance signalling may be audible using sounders. It is known traditionally as local signalling (bells only). The technique is often complemented with visual signalling using xenon strobes or beacons. Remote signalling if installed is performed to a point remote to the protected area by means of a hardwired, a wire-free link or a combination technique employing more than one technology.

Audible signalling is by means of electronic sounders, motorized sirens or electromagnetic or motorized bells. These are at most times supported by self-activating or self-contained bell modules and will be found enclosed in housings of various materials and styles. It should be understood that although the word 'bell box' is still used in the intruder industry this is a traditional term as the vast majority of sounders used within protective enclosures are indeed electronic solid state units.

The word sounder is in fact a time honoured expression but under the ENs they are classed within the terminology 'warning device' (WD).

Visual signalling devices range through xenon flashing lights to rotating beacons and will generally be configured on separate outputs to audible devices.

Remote signalling in all its forms is covered in this chapter. This signalling does at times include the alarm confirmation technologies invoked within the scope of the ACPO Security Systems Policy 2000 and DD 243, so this particular subject is detailed separately in Section 6.9. In addition to the three signalling methods we consider the electromagnetic relay as it can be used to interface heavy duty sounders and additional loads such as smoke screen protection.

How distance and climatic conditions affect the propagation of sound is to be appreciated as do those factors that determine the number of signalling devices that are to be specified.

In practical terms we have now reached the stage at which stand-alone security technologies have become increasingly rare, having been replaced by integrated systems and enhanced by the addition of communication networks.

This means that intruder alarms now make a much greater use of remote signalling and quiet signalling is becoming increasingly more prominent.

Interesting is the fact that under British Standards an 'external audible alarm was to sound. ... if its tamper detection operates'. However, Table 3 of EN50131 bans the use of a warning device if the system is unset when a tamper operates.

The ENs also brought about many changes in the way that signalling or 'notification' was to be carried out once the grade of system had been determined.

● To reflect the levels of security as denoted by the grading, special requirements now exist under the European standards for this notification.

In addition requirements were laid down for the communication device that reflects the maximum time and attempts that it takes it to communicate remotely.

There must also be measures taken to protect against equipment substitution including encryption techniques and the means of safeguarding the data that is being transmitted. This must prevent unauthorized reading of the transmitted data and an ability to detect efforts made towards the device being targeted for substitution.

The notification requirements are as follows:

GRADE 1
2 warning devices *or*
1 self powered warning device *or*
1 signalling system with 24 hour reporting (at least once per day)

GRADE 2
2 warning devices and 1 signalling system with 24 hour reporting *or*
1 self powered warning device and 1 signalling system with 24 hour reporting *or*
1 signalling system with 3 minute reporting plus encryption

Note. Grade 2X is recognized in the UK for non communicating systems

GRADE 3 *All signalling systems require encryption*
2 warning devices and 1 signalling system with 5 hour reporting *or*
1 self powered warning device and 1 signalling system with 5 hour reporting *or*
2 signalling systems with 5 hour reporting *or*
1 signalling system with 3 minute reporting

GRADE 4 *All signalling systems require encryption*
2 warning devices and 1 signalling system with 3 minute reporting *or*
1 self powered warning device and 1 signalling system with 3 minute reporting *or*
2 signalling systems with 3 minute and 5 hour reporting. Even faster communication devices with regular reporting every 20 seconds and methods to prevent unauthorized reading and modification of the transmitted data and the ability to detect the device being substituted can be specified.

As there are additional factors required of manufacturers of signalling equipment they will produce tabulations of their products. These tables can be used to list data such as security grades, environmental classifications, alarm transmission system values where applicable, and to provide other relevant information and comments. They will cover such products as warning devices and also signalling communication equipment. The evaluation will be based on agreed BSIA and security industry criteria.

The product tabulation could be typically:

Product	Security grade	Environmental class	ATS	Comments
Matrix	SG2	EC II	ATS 2	PD6662/EN50131

Speech diallers would not actually be given an EN compliance grade because periodic testing of the line to an ARC is not carried out and they have no reporting in the strictest sense. However they still have a tremendous use within the security industry due to the supplementary features they have in their own right so are covered at a later stage in this chapter.

Specifiers should be aware that installers are not able to claim compliance for their PD6662 alarms unless the alarm transmission equipment suppliers can provide them with declarations of the performance for the alarm transmission systems (ATS) in terms of the ENs. Reference can be made to EN50131-1-1: 1998.

ATS performance is summarized as being to ATS 1, ATS 2, ATS 3, ATS 4, ATS 5 or ATS 6 criteria with ATS 6 having the highest performance. It is possible to specify more than one ATS if the risk dictated this.

It will also be found that control equipment now tends to refer to outputs that are provided for warning devices/signalling equipment rather than using the traditional term 'trigger'.

6.1   Audible signalling devices

At a later stage, in Chapter 7, we go on to look at the various methods of securing external audible intruder alarm signalling devices to various surfaces. However, at this point we need to consider the different types of sounder that are available. We must first understand that a clear distinction applies in that sounders are classed as either:

- SAB – self-activating bell or self-alarm board;
- SCB – self-contained bell or board.

An SAB or SCB device is essentially an electronic module with an integral battery which is connected as an interface between the control panel and sounder.

SAB (Figure 6.1)

When the hold-off supply is applied from the control panel it energizes relay RL1, opening contact RL1/1 and disconnecting the internal battery

Figure 6.1   *SAB circuit*

from the actual sounder. The hold supply maintains the internal battery in a charged state. R1 is a resistor that limits the charge current when a NiCad battery is used as the standby source. If a lead–acid battery is used, then it is float charged by connecting it directly across the hold supply, and R1 is not needed.

In an alarm condition 0 V is applied from the control panel to the trigger terminal (negative trigger), and the trigger line carries the full sounder return current, although a proportion of the power is supplied via the hold supply. If for any reason the hold supply fails, RL1 becomes de-energized, and RL1/1 closes, connecting the internal battery to the sounder. Diode D1 is there to ensure that the relay is not latched by the battery. When a visual warning device is also used it will generally be activated by a separate input to enable it to operate after the sounder has been forced to cut off by the control panel trigger or an additional timer circuit on the SAB.

The SAB module may be purchased as a separate circuit, or forming a single unit with an audible warning device.

The terminal layout is typically that of Figure 6.1, with terminals +12 V and 0 V providing the hold-off supply. Unique terminals are provided for the sounder and strobe visual warning device.

SCB (Figure 6.2)

The sounder is powered by the on-board battery at all times when the system is in alarm. This also applies to any tamper condition or a loss of hold-off supply. This is achieved in a similar manner to the SAB method but in practice the SCB operates in normal alarm by removal of the trigger 0 V.

On occasions the installer will also note outputs that are programmable at the control panel for positive trigger. These are terminals held at 0 V but going positive in alarm, and therefore are used as the negative hold-off supply. Some panels achieve the same effect by programming the trigger terminal polarity. The trigger negative terminal at the control panel is thus

Figure 6.2 *Conventional SCB terminal layout*

connected to the hold-off 0 V terminal at the sounder, but the hold-off negative at the control panel is not connected and the option is set as SCB.

A further method of triggering an SCB is via an alarm output relay. With this technique the relay has voltage-free changeover contacts. The common contact C is connected to 0 V at the control panel, and the normally closed terminal is used to provide the negative hold supply and is effectively switched in alarm. The normally open contacts can provide the trigger for SAB operation.

This can also be achieved by using an additional relay module if the control panel does not have an alarm output relay.

Note that NiCad batteries are either 7.2 or 8.4 V, so if they are used for SCB operation they cannot deliver an audible output equivalent to that generated by the full 12 V from the control panel.

SCB devices make a smaller demand on the control panel than the SAB, and are therefore most suitable for multiple-unit operation.

Figure 6.3 shows this method of triggering via an alarm relay output. Relays are considered further in Section 6.3 since they can be used to great effect to perform a vast array of different trigger functions. See also Figure 6.4, Conventional SAB.

With the knowledge that the SCB places little emphasis on the capacity of the output from the control panel, unlike the SAB which only uses its sounder standby battery, in the event that its hold-off supply is destroyed, there remains one further aspect that the alarm engineer must appreciate with regard to many modern alarm control panels. The loops used in control panels, whether they are detection or tamper, are connected across two terminals. Within the panel they can be seen as a supply terminal and a sensing terminal.

Modern alarm systems use positive alarm loops of around 10 V and negative tamper loops connected to the 0 V line. The 0 V supply for the tamper loop is often derived from the negative hold-off supply so that the tamper can be sent over a single core.

Figure 6.3   *Alarm output relay*

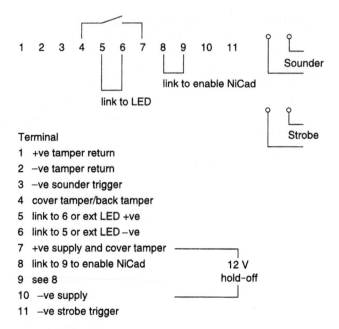

Figure 6.4   *Conventional SAB*

Consider certain modern control panels which incorporate a dedicated single tamper terminal for the external sounder unit. If an SCB mode is used with a single tamper return line then the tamper supply is removed each time the sounder operates under alarm or test. This will cause a tamper to be reported when no fault has occurred. If the relay method is used it is possible to wire a negative feed to the tamper return terminal at the control panel through separate relay contacts to substitute the lost negative signal for the duration of the sounder output. An extra diode is all that is needed to achieve this.

Warning devices

A wide choice of external audible warning devices is available. They will always be subject to attempts to force them from their mountings, to abuse from hammers and blow torches, and to drilling and to muffling by the injection of foam, and manufacturers attempt to protect these devices in various ways. The specific type selected for a system will be down to personal preference, cost and the likely risk of attack. The minimum standards for audible warning devices can be summarized:

- They must be of two notes. This can be continuous or 1 second on and off.
- There must be at least two fundamental frequencies from the range 300 Hz to 3 kHz, and must not allow confusion with sirens used by the emergency services.

- The housing must be totally enclosed and weatherproof. The protection should be that offered by 1.2 mm of steel, as a minimum.
- The device must generate a minimum 70 dB(A) mean sound level and 65 dB(A) in any one direction at 3 m with the protective enclosure in position.
- The sounder is to be self-powered by a battery source that provides either a 30 minute secondary cell or 2 hour primary cell alarm state duration. The maximum recharge on the secondary type is to be 24 hours. An anti-tamper circuit must be included.

In addition to these requirements the sounder should be wired so that it does not sound during normal opening and closing procedures. Also generally incorporated are integrated SAB or SCB cut-outs that will shut off the sounder in the event that the hold-off is destroyed, otherwise the electronics module cannot be controlled from the operator's control panel and will sound until the on-board battery is exhausted. The sounder itself is either a bell (electromagnetic or motorized), horn, cone or motorized siren.

At this point we should add that when detailing warning devices we do not consider internal sounders separately to external devices, but in practice piezoelectric or mylar cones tend to be used indoors. All sounders require good temperature resistance within the range −10 to 50°C at 10–95 per cent relative humidity. Indeed as we move up through the grades the manufacturer will offer the installer a warning device typically to the following specification:

- steel, louvreless, double-skinned with a sound level of 120 dB(A) when fully assembled and which draws power from its own on-board 12 V, 2.8 Ah(min) lead–acid battery;
- tamper protection will incorporate front screw, rear tamper or unit removal and drill detection;
- spare terminals to assist in the wiring of a strobe light;
- activation by a positive signal being removed or going to 0 V at the trigger input and remaining active for a time determined by the shorter of the trigger period of the panel and the on-board timer circuit;
- a pull-up resistor to provide a more secure system by producing an alarm if the trigger wire to the sounder assembly is cut.

Figure 6.5 shows the standard terminal layout. It can be noted that a terminal marked 'time out' is provided. This is a negative signal activated after the siren has concluded its period of time out. This negative signal remains until the unit is reset, and is suitable for driving a small relay, resistor and LED or buzzer up to 100 mA. An anti-tamper polarity select link is used to set the sounder to the control panel being used. Factory default values are generally negative.

The typical time-out signal connections are shown in Figure 6.6.

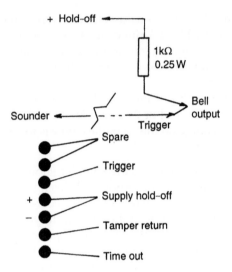

Figure 6.5   *SCB operation pull-up resistor*

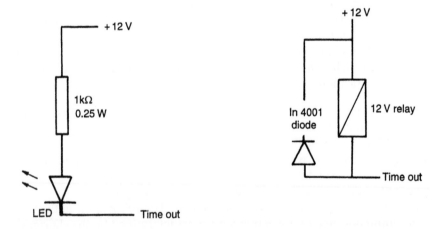

Figure 6.6   *Time-out connections*

Electromagnetic bell

This is the traditional sounder but has been largely replaced by the solid state electronic device. However, it will still be found in many older systems because it was capable of reliable operation over long periods of time if protected from the weather. The current is applied to a coil or solenoid and a steel plunger is magnetically attracted through the centre of the coil to strike a steel dome. As it strikes, a pair of contacts are opened, and the current is interrupted. The plunger is then returned to its rest position by a spring, its contacts reclose, the current is reapplied and it strikes again.

The distance travelled by the plunger determines the level of 'ring' between the dome and plunger: if the plunger is too close to the dome, the stroke is dampened by the plunger returning too soon, which results in a clattering rather than a ringing tone; if the plunger is too far away, it does not exert its full force on the dome, and the strokes occur at too slow a rate, producing a weak sound. The optimum distance is set by screw adjustment or by rotating an eccentric dome.

The design of the dome has an influence on the sound level and on the tone. A dished effect in the centre can give a harsher and louder sound. Domes are usually 152 or 203 mm in diameter. Although they may have the same coil and striking mechanism the larger version is expected to deliver 3 dB(a) more sound at 1 m. This type of bell is often found within a bell box, with the SAB or SCB module separately mounted in the enclosure.

Centrifugal bell (motorized)

This uses a small motor to rotate a striker at high speed which repeatedly strikes the inside of the dome, the direction and movement of the striker being influenced by centrifugal action. The current drawn may be marginally higher than for the plunger-type bell. The striker is once again set up by a screw to achieve the best ringing tone.

This type of bell is still extensively used, and has the SAB or SCB mounted under the dome to protect it. A polycarbonate cover then encloses the complete assembly, so that the device is totally self-contained.

Bell boxes. The original method of installing external warning devices was to purchase bell boxes with back-plates and then to place within them sounders with separate SAB (control panel powered) or SCB (self powered) modules. It is now more common to purchase high-performance bell boxes ready made up and claiming many options. These can include:

- Double-skin polycarbonate covers with high resistance to the weather claiming the IP55 standard.
- Steel shroud covers to prevent attacks by blowtorch.
- Louvreless enclosures to stop the injection of sound-deadening foam.
- Availability in a multitude of shapes and finishes for aesthetic reasons. To these can be added full-screen printing of logos to improve the overall effect.
- Drill protection. This is achieved by twin steel covers that electrically short the covers together when pierced by drilling.
- Fully encapsulated circuit boards to provide complete immunity to adverse weather conditions and with covers to protect the batteries.
- Angled or comfort LEDs, showing a charge condition, that act as visible deterrents.

- Variable cut-off timers to assure customers that extended audible signalling cannot be a problem.
- Anti-tamper protection front and rear to protect the integrated rechargeable batteries and the SAB/SCB.

We will see products designed to exceed the requirements of EN50131-4 Security Grade 2, Environmental Class III to include fully integrated sounder/strobes, a self-activating cut-off timer and twin status indication. The audible alarm will be capable of sounding in the event of an electrical change in the connections to the panel. Sounder and strobe monitoring on the bell box will indicate at any time if either the strobe wires are cut or removed which is achieved by the use of resistors that are fitted in the control panel at the installation stage. **Indeed, PD6662 states that warning devices used in a Grade 3 installation should have monitoring of the ring command (trigger wire monitoring).**

Low current draw is becoming an issue with warning devices and the ability to carry out diagnostic testing from a remote location in line with BSIA guidelines for remote maintenance.

Dual anti-tamper operation means that the circuitry is able to operate if either the cover is opened or the bell box is removed from its fixings.

The vast majority of purpose-made bell boxes also include integral strobe or xenon light units, which may be double or single or run around the perimeter of the cover to provide 360° illumination. Multifaceted windows on the cover can also be used to maximize light dispersion of the xenon light unit. There is still a need for separate visual warning devices, and we consider these in the next section.

Other types of sounder

For these types of audible warning devices the terms 'sounder' and 'siren' can be used interchangeably.

The sounders themselves may be driven by electronics (solid state electronic horns, piezoelectric sirens or mylar cones (single or twin)) or mechanical sirens. Solar powered sounders and wire-free devices are also available.

Solid state electronic horns. These comprise an oscillator amplifier and horn-type loudspeaker both contained within the same unit. They are popular as external sounders because not only can they deliver 120 dB(A) drawing only some 400 mA with no significant inrush current but they are also inexpensive. The horn is re-entrant in that it is folded within itself to save space, but it is a highly directional device and needs to be directed towards the most effective area. The frequency of the generated tone or its pitch is found to vary between models. The ear is most sensitive to frequencies between 2 and 4 kHz, but tones in this region and above are attenuated at distances beyond 50 m. At long distances the losses cancel the advantage of frequencies in this band, but those slightly below it (of the order of 800 Hz to 1 kHz) are very effective, their tone carrying better at distance, and for this reason are the most used.

Piezoelectric sirens. Also extremely popular with its high-output and high-pitched tone and with low current consumption, these can be used confidently for both external and indoor applications. The principal element of these is a ceramic piezoelectric transducer which generates an audible tone and frequency when it is energized by a peak square wave driven direct by a low power consumption IC. The transducer is mounted rigidly in its housing, and this siren type is extremely reliable in operation and can withstand severe environmental conditions. Because of the high pitch of piezoelectric sirens, the tone does not carry quite so well over long distances; however, these devices are excellent for short-range external duties.

Mylar cone speakers. These are often found where multiple tones are needed. They may be mounted on a square or round chassis and range in size and appearance from circular 50 mm to square 100 mm devices, with an output up to 100 dB(A) at 1 m. The power consumption ranges from 0.5 to 15 W. They comprise a cone speaker made of Mylar, which is a transparent film that is unaffected by moisture. Ferrite magnets with a plastic centre dome combined with a card gasket around the front make them weatherproof from this direction. They are not affected by humidity, and are highly reliable. They are capable of good, low-frequency, long-range transmission, and can resist foam attack if mounted as rear facing.

Mechanical sirens. These consist of an electric motor which drives an impeller that in turn forces air through vents in the casing in such a way as to create a loud sound. The unit is semi-enclosed and can be used confidently in outdoor but not indoor applications. They can deliver 120 dB(A) at 1 m, but their running current may well be of the order of 1 A, and the inrush considerably more. Many mains sounders operate using the same technique and can be used as additional signalling equipment alongside 12 V sounders if powered via a relay.

Propagation of sound

Before the positioning of any type of sounder to standard or specification, it is important to have an understanding of noise transmission or propagation.

Loudness follows a logarithmic rather than a linear scale, so a sound measuring twice the level of another will not seem twice as loud.

Consequently, a sound that is twice as loud as another would in practice be many times the level of the other. To specify the aural effect of the loudness of a sound we must use units that compare the measured sound to a standard reference level in a logarithmic manner. The reference level is the human threshold of hearing in a healthy young adult – a sound pressure of 20 Pa. Other sound levels are expressed as a ratio of this, which is thus given the value 0 decibels (dB).

Anything over 80 dB(A) is considered loud; 100 dB(A) is classed as unpleasant; and 130 dB(A) is the threshold of pain, and exposure to this can cause permanent hearing damage in a matter of minutes. The decibel can be used to specify a difference in level between two sound sources or the difference in level at different distances of the same source.

Consider some typical everyday sound levels:

Inside a room in a quiet neighbourhood	20 dB(A)
Ticking watch	30 dB(A)
Whisper at 1 m	45 dB(A)
Vacuum cleaner at 1 m	75 dB(A)
Average disco level	100 dB(A)
Pneumatic drill at 1 m	105 dB(A)

Sound levels decrease with distance, and most sounders are specified at 1 m. Sound pressure for most sources drops in proportion to distance, so at twice the distance the pressure is a half, and at three times it is a third.

Some common decibel ratios are:

6 dB(A) × 2
10 dB(A) × 3
12 dB(A) × 4
18 dB(A) × 8
20 dB(A) × 10

The dB(A) value for a ratio of 1:3 is 10 dB(A), therefore a sounder rated at 90 dB(A) at 1 m is the same as one rated 80 dB(A) at 3 m.

A sounder at a height of 6 m is only one-sixth of its quoted rating at 1 m, or 16 dB(A) less.

Placement of a sounder is therefore important. Consider premises where the distance from the front to the back is, say, 20 m. This means that the sound could drop to a twentieth or 26 dB(A) of its rated value if the distance between the sounder and a listener is at a maximum. The minimum sound level at which a sounder is audible above moderate ambient noise is 60 dB(A).

Table 6.1 illustrates the distances at which a given level at 1 m from the source decreases to 60 or 70 dB(A); thus, the distance over which a given sound volume will be maintained from a specified source can be found. Alternatively, the required source level to obtain a given volume at a known

Table 6.1   *Distances related to dB(A) levels*

Sound level at 1 m (dB(A))	Distance for 70 dB(A) (m)	Distance for 60 dB(A) (m)
83	4.4	14
86	6.3	20
90	10	32
93	14	45
96	20	64
100	32	100
103	45	144
106	64	203
110	102	322
113	144	456
116	204	645
120	324	1024

distance can be determined. From this table it can be seen that sound levels decrease with distance from the source generally at a rate of 6 dB(A) for a doubling of the distance from a point source.

Theoretical dB(A) levels such as those in Table 6.1 must of necessity be modified by obstructions and absorbent surfaces. Also, levels in still air will be affected considerably by any breeze, increasing the range in the direction in which it is blowing and decreasing it in all others. The normal temperature gradient whereby air gets cooler with height causes sound to refract upwards and thus also reducing the range. There can be temperature inversion when air above the ground is warmer than at ground level. This causes sound waves to be bent downwards and increases the range at which sounds can be heard.

We can conclude by saying that the installer must select the best sounder type relative to the environment and this must be as inaccessible as is possible in its mounting position. It must be fixed to a solid structure of concrete, brick or hardwood if sound energy is not to be lost in vibrating the building structure. The area is best free from high, dense hedges or trees that will attenuate the sound, and an audibility test should be carried out after installation, selecting the worst attenuation path. It is important to remember that sound reduces by 6 dB(A) for every doubling of the initial measurement distance on flat, open sites, and by up to a further 10–12 dB(A) in areas of vegetation.

Considerations

For a brick wall it is industry practice to mount the warning device with a minimum of three No.10 screws penetrating the brick (not the mortar or facing) by at least 40 mm and screwed into suitable wall plugs. For thin structures and metal or wooden clad structures or asbestos it is necessary to use bolts which penetrate the material and these are to be used in conjunction with a back-plate. The device is to be 3 m or more above ground level. It is to be appropriate to the premises, articles and operations.

Of course devices are best with a means of detecting drilling and penetration by foams. The importance of the external warning device is not to be underestimated, especially with the huge range of devices available at the present time and at such competitive rates.

6.2   Visual signalling devices

Some form of illumination or visual indication linked to an alarm is always desirable for better identification of an alarm source.

Many new-generation self-contained bell boxes contain integrated xenon flashing lights, and there is a good selection of xenon flasher units/alarm beacons available for mounting on the outside of other bell boxes. Alternatively, they can be mounted on the inside of translucent bell boxes.

The light source used is the xenon tube, which is a glass tube filled with xenon gas that emits a brilliant flash when a high voltage is momentarily discharged through it. This high voltage is produced by an integral associated circuit in the beacon.

The brightness is determined by the electrical energy discharged, and is rated in joules. The joule is a power of 1 W sustained for 1 second, but as the flashes last for only a fraction of a second the light intensity is equivalent to that of a high-wattage filament lamp. A typical value is 5 J.

A 5 J device can be visible for several kilometres, depending on the background ambient illumination.

A typical low-profile sealed unit available in a number of different colours can have a flash rate of three per second and have a supply current of 150 mA with a rated life of the order of 1 million flashes. These units, which are normally fixed using 4–6 mm diameter bolts, tend to have bodies of ABS plastic, with the cover manufactured from translucent acrylic marked with Fresnel rings and striated sides to maximize light dispersal.

The low-profile unit is, of course, for use where space is at a premium, but for maximum effect high-dome beacon assemblies are advocated. These are industrial grade and of continuously rated rotating mirror design with a choice of lens colours and motor voltages. A parabolic reflector circles a stationary lamp, giving a powerful sweeping beam of light reflecting off surrounding objects and illuminating any surroundings. These are suitable for applications which do not experience high levels of vibration or shock. They have high environmental protection (IP65 typical) and comprise a quiet, direct-drive motor (no drive belt) within a high-impact ultraviolet-stabilized polycarbonate enclosure. Depending on the motor size, these units may need to be controlled through a relay.

High-power xenon beacons of a similar size, of the order of 200 mm or 150 mm in diameter, are also used in applications where high levels of light transmission are needed. Unlike the rotating beacon, these allow horizontal and vertical 360° viewing, using two different types of exciter:

- *Single hit.* This uses all the available flash energy in one discharge to offer a very intense flash event for maximum penetrative/distance effect.
- *Double hit.* The flash energy is shared by a primary flash (66%) followed by a secondary flash (34%) a short time later, giving a very insistent hazard warning owing to the response characteristics of the eye.

When it is desired to illuminate an area, in addition to energizing a visual alarm signalling device, this is accomplished by means of a 12 V coil relay switching the supply for the lighting through the main contacts.

6.3 Electromagnetic relays: applications

In Section 2.3 we met the electromagnetic relay, which is activated when a coil is energized. Within the security industry in general these devices are used widely as a means of interfacing systems or to switch greater loads through the main contacts of the device using additional power supplies.

Relays can be placed into two categories for our purposes. In the main category there are four particular subclassifications of relay, namely:

- general-purpose relays;
- plastic sealed relays;
- hermetically sealed relays;
- power relays.

The second category covers special-purpose relays, of which there are two subclassifications:

- latching relays;
- ratchet/stepping relays.

These types are easily summarized:

- *General-purpose relays.* The contacts turn on instantaneously when the coil is energized and off when de-energized.
- *Plastic sealed relays.* The mechanism is encapsulated in a plastic case with the terminals and terminal block sealed by epoxy resin.
- *Hermetically sealed relays.* The internal mechanisms are completely sealed from the external atmosphere by a metal case and metal terminal block.
- *Power relays.* These are intended to switch heavy loads.

- *Latching relays.* The contacts of these relays are magnetically or mechanically locked in either the energized or de-energized position until a reset signal is applied.
- *Ratchet/stepping relays.* The contacts of these relays alternately turn on and off or sequentially operate when a pulse signal is applied.

Having selected a category of relay, we next consider the supply voltage needed to operate the coil. Two types of relay coils are available: AC or DC operated (Table 6.2).

Generally, contact ratings are given based on a purely resistive load, and a highly inductive load with a form and contact material to aid in best selecting the optimum relay with a certain needed service life. In the intruder alarm sector the type of load being switched will be essentially resistive so there will be little arcing using a DC power supply with a high current. Nevertheless, there are recommended ways of employing contact protection circuits and also to suppress noise as the relay contacts open. The methods of surge suppression vary and use capacitors, diodes, zener diodes or varistors, but these will rarely need be used in the alarm sector if the correct voltage relay is selected with the required contact materials.

The features of these contacts are easily related to the load current. Table 6.3 details the change in material with load current.

With this information to hand the student can then establish applications for relays such as the following examples:

- For signalling in an alarm condition the applied 12 V can be used to energize the relay coil to bring on mains lighting through the relay contacts switching the phase only in a parallel circuit to the existing light switches. By a similar technique, high-output 240 V sirens can be driven.
- Mains coil relays can also be energized from the 240 V supply to select an opening or closing low-voltage loop.
- High-output xenon strobes or non-solid state mechanical sounders can also be driven through the relay contacts by use of an additional power supply.
- Interfaces can be made with other security systems by energizing relays to open or close loops or provide inputs to start remote signalling equipment.

The applications that will be encountered can therefore range through signal control to power drive with momentary functions to the latching of circuits and inputs until further pulses are offered.

Table 6.2 *Relay coil voltages*

AC coil	6 V	12 V	24 V	–	50 V	220/240 V
DC coil	6 V	12 V	24 V	48 V	–	220/240 V

Table 6.3 Contact material selection for a relay

Low load current → High load current

PGS alloy (platinum–gold–silver)	AgPd (silver–palladium)	Ag (silver)	AgCdO (silver–cadmium oxide)	AgNi (silver–nickel)	AgSnIn (silver–tin–indium)	AgW (silver–tungsten)
High resistance to corrosion. Mainly used in minute current circuit	High resistance to corrosion and sulphur	Highest conductance and thermal conductance of all metals. Low contact resistance	High conductance and low contact resistance like Ag with excellent resistance to metal deposition	Rivals Ag in terms of conductance. Excellent resistance to arcing	Excellent resistance to metal deposition and wear	High hardness and melting point. Excellent resistance to arcing metal deposition and transfer, but high contact resistance and poor environmental durability

6.4 Remote signalling: the alarm receiving centre (ARC)

Traditionally the role of the ARC has been the watchful guardian of monitored security and fire safety services. However, they are now obliged to carry out additional services to include the care industry whilst also showing an interest in the monitoring of commercial and industrial processes. Therefore the ARC is not only checking signals from time honoured digital communicators but is now also dealing with PC-based systems that are providing high levels of information and data. They must now cope with new-generation communication devices alongside remote video surveillance and various alarm technologies.

With respect to intruder alarms, we can give an indication of some of the main and additional services in Figure 6.7.

- **redcare.** The British Telecom (BT) alarm signalling network. Secure and reliable it is a continuously monitored analogue line and can alert the ARC immediately that a line is cut or damaged. redcare ISDN is the digital option capable of detecting high volumes of different alarm states and allowing information to be delivered to the ARC in text format. It uses the ISDN2e or Business Highway to transmit alarm activations. A great deal of different redcare systems exist to satisfy different customer requirements. For example, redcare assure is a form that combines dual path IP alarm signalling with video monitoring technology.

Figure 6.7 *ARC activities*

- **redcareGSM.** Capable of relaying signals after landline attacks as it uses the cellular radio network to deliver back-up transmissions
- **Dual communications.** The use of digital communicators plus radio links to give a secure combination. This term can also relate to any two different communication links used in tandem to provide the necessary additional transmission medium
- **Radio.** A specific radio link with polling and immunity to line fault false alarms
- **Digital communicator.** The basic form of signalling between the protected premises and the ARC
- **Point ID.** An enhanced digital communication technique with alarm reporting data
- **Alarm confirmation.** Audible, visual or sequentially confirmed alarm signals.

The role of the ARC has changed enormously over the years to deal with a complex role of additional services and duties. Therefore it will be found to embrace guard response and paging with message handling plus web based products that are redefining the way installers manage and control customer data bases on line. Video monitoring and digital recording are always included to ensure total security of assets around the clock. A comprehensive account system with customer details and site information makes the management process efficient with different levels of access rights assigned to each station operator.

The modern ARC will use all of the latest transmission products to include PSTN, ISDN, IP, GSM, GPRS and X25 through to network and ADSL internet monitoring. European Standards covering system design, installation and maintenance and monitoring are the key to a new strategy for the ARC. It has a commitment to quality with approval by particular governing bodies to BS 5979 and ISO 9000.

BS 5979 was actually rewritten in 2000 to take into account remote monitoring and refers to high-security alarm centres. It is to this standard that the control centre must be constructed and secured. For those control rooms that are not required to comply with BS 5979 a further standard exists, namely BS 7499 *Construction and Operation of a Control Room*, and this can be used for guidance. A further standard BS 7958 which relates to control room operations details the operational procedures and practices for those carrying out such activities. The vetting of staff is an important part of a manager's role and BS 7858 *Security Screening of Personnel in a Security Environment* can be used as a template for vetting procedures to ensure all relevant staff are screened to a minimum standard.

As we know, activations from monitored systems must be routed through the ARC for verification prior to being passed to the relevant authority for police or fire response. BS 8418 *Remote CCTV Monitoring: Detector Activated* aims to minimize false alarms by using detectors to trigger CCTV activations, with images being transmitted to remote stations. This scheme calls upon a wide range of standards and requires the facility to 'audio challenge' a potential intruder. A unique reference number for the installation is required to initiate a police response. An immediate level 1 police response to detector activated CCTV is only available when the system complies with BS 8418 and is therefore able to qualify for a Unique Reference Number. It follows that a system inspected to this standard not only ensures quality, but is also a means to determine that an incident on the premises becomes a police priority.

The National Security Inspectorate, an independent non profit approvals body providing inspection services for the security and fire industries, has an ARC Gold scheme to which ARCs can be approved. This status confirms that the ARC meets the industry specific ISO 9000 Quality Management System standard and operates formal inspected internal processes to give high levels of customer assurance. Within the terms of the approval status the ARC must also be able to demonstrate a long term track record of performance and reliability and stability with a high level of insurance specification.

The Security Systems and Alarms Inspection Board also have a formal Alarm Receiving Centres monitoring stations certification scheme. Certification in this instance is to verify that the monitoring station demonstrates compliance with the ACPO policies on police response to security systems.

The construction and operation of the ARC must be to BS EN ISO 9000 and a category of BS 5979 with further approval by the British Security Industry Association (BSIA), the Association of British Insurers (ABI) and the Loss Prevention Council (LPC).

Strategies to improve the efficiency of intruder alarms include an aim to improve the quality of information to the ARC and the quality of information subsequently relayed to the police from the ARC.

Therefore it follows that the monitoring of the Alarm Receiving Centre is incredibly diverse and now extends beyond intruder concepts even though this remains a core role.

Of course we will come across other terms in addition to those of the alarm receiving centre.

These include the remote video response centre (RVRC), control rooms and CCTV monitoring centres. These all play a central and critical role in the security industry.

The RVRC reacts to the triggering of detectors linked to a CCTV system. Images of the incident are transmitted to the RVRC for verification and response. If intervention is required police response may be initiated using the protected sites URN.

RVRC installations are also required to have an 'audio challenge' facility to deter potential intruders. RVRC operators reactively monitoring private property (and therefore by default) ARC operators visually confirming

alarm activity need to be licensed. To apply for a frontline license an operator has to pass an industry exam and undergo formal training.

The private security industry is a massive provider of services and is
increasingly being used as a valued member of the 'wider police family'
with a key role to play in the prevention of crime and terrorism. For this
reason we will surely find the application of all monitoring stations becoming increasingly and more diversely employed.

The student may also come across the term 'satellite station'. Satellite stations are not permanently manned but are normally unoccupied rooms
within a building that has its location in reasonable proximity to a number of
protected premises. The signalling systems of all these protected premises is
connected to the satellite station by means of a single dedicated telephone
line. This line will generally be a private line. The satellite will itself have
multiplex facilities to transfer all of the incoming signals into a single outgoing network or trunk line. It is this trunk line which connects the satellite
station to the permanently manned central station, which is normally within
the same police region (Figure 6.8). This provides certain advantages:

- the lines linking the protected premises to the satellite station are
  monitored;
- multilines between the satellite and the central station can be utilized in
  the event of failure of the normal line;
- due to line sharing of the multiplexed trunk line there is a cost saving.

However, other aspects should be noted:

- generally for premises in densely populated areas;
- dependent on the automatic receiving equipment rather than manning
  personnel.

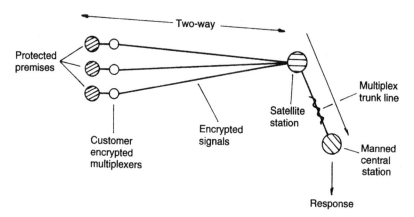

Figure 6.8   *Satellite station*

6.5   Remote signalling: telephone lines

Having established that satellites are mainly for cities and larger towns, we must consider more rural applications where the telephone company exchange can be seen as a substitute for the satellite. In these applications, instead of utilizing a multiplexing unit within a satellite, each site will transmit a unique multiplexed signal. When received at the exchange the various signals are run into a single shared line which is connected to the central station (Figure 6.9).

Line types

We should now become familiar with the main types of line used to transmit alarm signals. The familiar public switched telephone network (PSTN) is of analogue form with the subscribers' line consisting of a single pair of wires connecting the subscriber's premises to the local exchange. Such a simple connection facility carries the voice frequency communication in both directions in the form of speech, of modem or facsimile tone or of other tones. Furthermore, it carries signalling information both from the subscriber to the exchange and vice versa. The integrated services digital network (ISDN) makes use of a wider bandwidth and higher data transfer rate so is better suited for transmitting data at a higher speed.

The actual routes/directions of the particular lines that have been installed are to be understood because of the different levels of security that they provide. These are:

- the direct line (individual);
- the shared direct line;

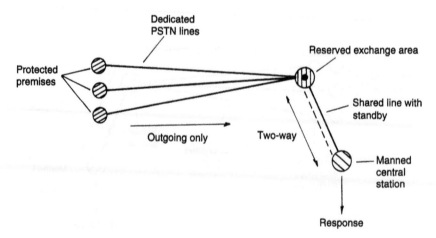

Figure 6.9   *Rural exchange using PSTN*

- the indirect non-dedicated exchange line;
- the dedicated exchange line.

The direct line (individual)

This is a single line that is hardwired, running from the protected premises and normally routed direct to the police, although it can also be defined as routed direct to a central station which can send out a response force (Figure 6.10). The main criteria is that it is direct and is not linked with an exchange. It is designed to be used in conjunction with a communication device that can constantly transmit a signal that is coded and that can be monitored for line failure.

The individual direct line is always advocated for high-security applications including bank and government installations. It can also be used in conjunction with communicators that have random code sequences to ensure that the code cannot be analysed.

The clear advantage of the individual direct line is that it is dedicated and not in any way shared, so it is not governed by any other signals. It also has a direct action without any intermediate persons involved. Equally, there is an immediate warning to give the fastest available response. There is also no switching of the line or any manual action needed. It has high installation and maintenance costs which tends to preclude its use for mid-security risks.

This system also takes up space at the monitoring point, and can have an excessive cost in working hours particularly if there is any false alarm risk.

Other disadvantages are that long line distances create even greater expense and that only basic signals can be transmitted.

In certain cases it can be of benefit to employ a reassurance capability within the protected premises to monitor the line state and receiver although this does add further to the expense of the system.

The shared direct line

This differs from the individual direct line only in that the single, hardwired cable also carries signals from a number of other protected premises which

Figure 6.10   *Direct line (individual)*

Figure 6.11   *Shared direct line*

Figure 6.12   *Indirect non-dedicated exchange line*

tend to be in the same general area (Figure 6.11). These signals are multiplexed onto the line and are carried simultaneously to the central station.

The advantage of the sharing of a direct line clearly lies in spreading of the cost among the various protected premises, but it is disadvantaged in its lower inherent security to that of the individual line. There is clearly a maximum coding that can be effected, and it is more effective when all the premises are in close proximity to each other.

The indirect non-dedicated exchange line

This is a common method of connecting dialling equipment to central stations via a non-dedicated single-direction PSTN line (Figure 6.12). It is indirect since it is directed via a BT telephone exchange, so a switching process exists.

This type of line is not dedicated to the alarm signal and can also be used as an outgoing telephone line, although it must of necessity be ex-directory.

With this format the dialling equipment is connected to the existing ex-directory 'call-out only' line, which if it is to give even a limited amount of security should not be accessible to an intruder, and be of the underground transmission variety.

Figure 6.13   *Dedicated exchange line*

Advantages obviously exist in cheap line rental and user costs since it may additionally be used for standard telephone calls when the intruder alarm is not activated.

It is at a disadvantage in that it offers no monitoring of interference by an intruder and cannot be checked readily for false alarms, but it still remains in wide use for low- to mid-security risk applications.

The dedicated exchange line

The fourth type of line is the dedicated exchange line (Figure 6.13). This is similar to the indirect non-dedicated exchange line but is dedicated to the alarm only and not used for any other outgoing calls. Like the previous type it is also routed through an exchange, but because it is dedicated, better security of the line to the exchange is obtained, but it can also be considered indirect. It is not cost-effective for many installations because of the restriction that the line must be dedicated to the alarm signal only.

Having gained an understanding of the types of line we may conclude that the indirect system is a line that is routed through a telephone exchange and then linked to a central station manned 24 hours a day. In earlier times such lines would have been linked to the police station, but the collection of data from alarm signalling equipment is now very much more in the hands of the private sector.

This indirect transmission of activations to the police (in that it is now directed through a central station) does slow down the process of following up the cause of an alarm condition but it is the only method now advocated. Central stations can clearly interrogate systems before contacting the police and hence ensure manpower is more effectively employed and false alarm conditions filtered out.

The next topic of study must be that of assessing multiple transmissions over a single path.

Multiplexing

This is the simultaneous transmission of multiple signals over a single path with recognition of each signal being made and uniquely identified by the receiver. There are two recognized types, time division multiplexing (TDM) and frequency division multiplexing (FDM), and both are used for hardwired remote signalling systems.

In multiplexing each signal is different from that of its neighbour, being coded to a specific circuit or device that it addresses or interrogates when the signal is asked to report its status.

Multiplexing allows the adoption of shared transmission paths rather than having to use dedicated lines directly connected to central stations with their inherent high costs. Shared lines are clearly less expensive in terms of rental, and few premises actually need dedicated lines since alternative path switching and fail-safe alarms make shared lines satisfactorily secure.

The use of shared lines also permits the use of higher transmission and data rate lines, which would be expensive to purchase on an individual basis.

Frequency division multiplexing (FDM)

This is a true simultaneous signalling transmission method (Figure 6.14). Rather than all of the premises receiving and transmitting in turn, they simultaneously transmit over a shared line but these transmissions are individually identified by the exact frequency and waveform of the dedicated signals.

Only information sent over telephone lines and by microwaves can use FDM, and even then the signals must be within specific parameters and limits known as the system bandwidth. This bandwidth is narrow, so very few discreet frequencies can be carried on the same line.

Time division multiplexing (TDM)

This is not simultaneous transmission in the same sense as FDM because even though the frequencies and tones may be similar to FDM they are

Figure 6.14   *Frequency division multiplexing*

separated by time. The central station transmitter sends a series of coded signals over the shared line in a cyclic pattern to the premises being protected, and this is repeated on a regular basis (Figure 6.15). All premises must then answer the signal by means of its own encoded response transmission within a specified time scale or similar response cycle.

Consider an entire cycle of transmission for a number of premises taking N seconds before being repeated. If one of the protected premises does not respond, then an alarm is generated at the central station. If the alarm signal is still not responded to after a further N seconds then an alarm is confirmed, although the number of failure cycles will also be governed by the equipment and line length. This type of system is seen as two-way interactive and continuous in being monitored. It has great potential, and is superior to normal unidirectional alarm signalling.

Alarm control panels use coded version of TDM to communicate with the central station. Using a pulse code modulator (PCM) or stream burst modulator (SBM) the signals are coded, which improves security of the signals received from individual protected premises. These signals form a binary code or as a pattern of particular frequencies. Each building will have a designated address code together with other features such as remote management and maintenance function codes. These will cover requirements such as:

- property address code;
- arm system code;

Figure 6.15 *Time division multiplexing*

- disarm system code;
- test system code.

6.6   Remote signalling systems and connections

In the first instance we need to consider the logic of remote signalling for an intruder alarm as illustrated in Figure 6.16.

In practice the actual configuration of the alarm transmission system is governed by the operational features and may include the use of more than one alarm transmission system and perhaps more than one ARC. It is also possible for a supervised premises to be connected by more than one form of transmission system.

Transmission systems under the EN standards are common with intruder, fire and access control networks etc. so are identified under EN50136 and not uniquely within EN50131. The transmission characteristics, system classifications and other criteria can be found listed within the following EN50136 standards:

- **EN50136-1-1** General requirements
- **EN50136-1-2** Requirements for systems using dedicated paths
- **EN50136-1-3** Requirements for systems with digital communicators using the public switched telephone network
- **EN50136-1-4** Requirements for systems with voice communicators using the public switched telephone network.

EN50136-2-1 is in three parts as follows:

- **EN50136-2-1** General requirements for alarm transmission equipment (ATE)
- **EN50136-2-2** Requirements for equipment used in systems using dedicated alarm paths
- **EN50136-2-3** Requirements for equipment used in systems with digital communicators using the public switched telephone network.

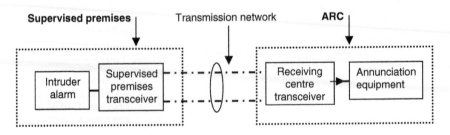

Figure 6.16   *Alarm transmission system*

The role of remote signalling has changed enormously over the last few years, however if we cast our minds back a while it would be noted that the original method of connecting alarms to the police was with automatic dialling equipment by direct line. In order to cure the faults attributed to automatic dialling equipment in the early days, other forms of signal transmission were then introduced. This all coincided with the advent of the Alarm Receiving Centre.

Alarm by carrier (ABC) was one of the early techniques.

Alarm by carrier

This uses a normal telephone line to the local exchange that is monitored permanently for failures and faults and for the alarm signal (Figure 6.17). These signals are 'carried' on the top of local audio telephone line signals which simultaneously use frequency division multiplexing. Any faults are reported to a central reporting station by the local exchange. Alarm signals are collected on a dedicated part of the exchange known as a local

1 reports alarm to 2
2 reports alarm to 3
3 checks and acknowledges signal. It then sends to 4
If faults occur anywhere 4 reports to 5. Alarms always take precedence over faults.

Figure 6.17   *ABC system*

processor. The alarm signals are then transmitted to the area exchange processor, which in turn routes calls to the appropriate police force after having rechecked the line and message content. Having received the signal, the police would then send an acknowledgement back through the system.

This transmission method offers the benefits of monitoring for faults and failures. It also offers a semi-dedicated line without recourse to additional line rental charges or installation costs.

Remote signalling communication practices

Advances in technology came along quickly as the intruder alarm industry expanded over the years and even mainstream units of control equipment provided a facility to connect to communicators. An alternative was to connect to redcare equipment via a subscribers' terminal unit known as a STU.

Internal communicators on the main PCB of intruder alarm equipment became commonplace. In addition, the control unit could often be connected to a separate communication device using a wiring harness connecting to interface pins on the main PCB.

The internal communicator could be used for:

- Transmitting alarm signals to alarm receiving equipment at a monitoring station using a number of different formats
- Connecting to a PC based at a remote engineering centre equipped with specific software to upload or download the system parameters (including the event log and diagnostics). It could also set and unset the alarm system and carry out additional special functions.

From all this a huge variety of communication equipment became available to satisfy the industry needs. As an example snap in digi-modems allowed a 'bells only' panel to transform to a fully communicating system using fast format, contact ID or Security Industry Association (SIA) protocols for alerting a central station. SMS text messages for a variety of events could be sent to give users peace of mind, and full and secure control of the system could be achieved via a PC running unique software. This allowed remote setting, printing logs, observing events and even changing user codes. In all instances the connection details are provided in the installation manuals which are supplied by the manufacturer. However, there are a number of essential considerations that hold good in all instances. Communicators can be connected directly to an ex-directory line used exclusively for alarm communications.

If an exclusive line is unavailable and other apparatus such as an answer phone is connected to the line, the communicator should provide a series connection with line divert for the other apparatus. When triggered the communicator is designed to seize the line and to disconnect the other apparatus while the communicator remains operational. When the communicator has completed its operation it will then reconnect the other apparatus.

Three way calling facilities, when provided, ensured that a communicator can use a shared telephone line. The local exchange must provide the Three Way Calling service on the subscriber's line.

Statutory information/safety notices

It is to be understood that statutory information and also safety notices exist in reference to the suitability of connections to the appropriate networks, such as direct exchange lines (PSTN) supporting DTMF tone dialling and PABX exchanges with or without secondary proceed indication.

Although equipment may be manufactured to meet all European Economic Area telecommunication networks requirements, there may be differences between the individual PSTNs in the various countries. Therefore equipment should be verified as suitable for the exact location it is to be installed. It can be said that Council Decision 98/482/EC for pan-European single terminal connection to the public switched telephone network contains approval criteria.

The BSIA has also warned installers of potential incompatibility between ADSL broadband and monitored intruder alarms when fitted in the same premises. ADSL broadband consists of an analogue line transmitting down one cable. If an auto dialler or digital communicator is used for signalling over broadband to an ARC, these signals may be corrupted or blocked as a result of the digital signals causing noise on the lower frequency analogue signals as the signals may overlap. Resolving this issue is a matter of fitting a Digital Subscriber Line (DSL) microfilter as it removes the need for splitters or the addition of an extra line.

Communication with mobile phones also became a futuristic issue with intruder alarms and the new GSM version of the Freewave wire-free system. This enabled the ability to text up to a given number of telephones both mobile and landline with customized messages to report such things as alarm, and entry or exit events. It could be controlled via a series of instructions sent by text or DTMF and could also communicate with an ARC via GPRS or direct dial by contact ID.

There are always certain considerations for any communication devices or remote signalling medium. When becoming involved with communication signalling devices note that terminals on the main PCB control unit may be described as safety extra low voltage (SELV), according to EN60950. Terminals on the communication device may be described as SELV or as telecommunications network voltages (TNV), also according to EN60950.

In order to comply with safety regulations:

> *TNV terminals should only be connected to PSTN or other circuits designated as TNV circuits.*
>
> *SELV terminals must only be connected to other circuits designated as SELV circuits.*

Connection of the telephone line

Direct connection of, or the interconnection via other apparatus to the communicator are capable of producing hazardous voltages on the telephone network so there are specific procedures to follow.

We can go through the more general connection methods in their most simple terms as follows:

In some instances the telephone line may be already terminated with a suitable lead and plug so that it can be inserted into an RJ11 connector on the PCB on some units of built-in communicator equipment.

If this is not the case it is normal to use approved cable (1/05 mm CW1308) and connect it to the A and B terminals on the communicator and to the corresponding terminals on the BT master box.

If other apparatus is required to share the telephone line with the communicator (series apparatus), it is normal to connect the main apparatus to the series switched line connections that are provided – these are generally identified as A1 and B1.

The A and B lines must be connected correctly for polarity. The communicator will monitor the line for ringing tones. See Figure 6.18.

As previously mentioned redcare has a prominent role to play in the intruder industry and is both well established and innovative.

Figure 6.18   *Telephone line connections with divert*

Communicating alarm response equipment (redcare)

redcare provides links between the alarm system and the central station (Figure 6.19). It is a network that both monitors and checks the system at all times, so if the alarm is activated or the line cut an appropriate signal is sent to the monitoring station. It also works in the event of the line being engaged. Equally, there is no need for a new separate line unless the line is also used for data transmission such as by fax or modem.

Of all calls to the emergency services, a high proportion are false alarms, but redcare automatically checks that the fault is not an electrical interference pulse. The system may be reset remotely by the central station operator. The transmitter or premises alarm communication equipment (PACE) offers up to eight channels; for example, fire, personal attack, intruder and opening/closing.

The service transmits its emergency or fault signals through two routes. Even in the event of a network fault there is no factor to prevent the signal being received. Signals are encoded in a unique way for the premises, and are only recognized by the redcare equipment.

Despite the signalling systems that we have encountered, we should not forget that the public switched telephone network is still well established for alarm signalling. Indeed, among its newer applications is picture transmission using modern high-speed error correcting modems (as an alternative, the integrated services digital network can be employed as this is a true digital circuit, and in practice, better able to relay images). Some insight is given in Section 6.9 into picture transmission and how it is adopted as a verification method, so now we shall consider other methods

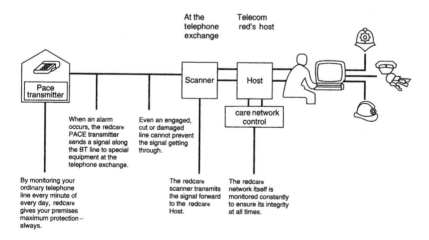

Figure 6.19 *redcare*

of remote signalling. Remote signalling must essentially include two basic elements:

- a means of communication with a response force;
- a network through which the message can be transmitted.

redcare

In keeping with these elements redcare has developed a range of products and services to fulfil the needs of the security industry at all of the different levels. It extends to redcare assure systems that combine dual path Internet protocol (IP) alarm signalling with video monitoring technologies. Therefore alarm signalling may be over IP with video recording triggered by activated intruder or hold up alarms.

The range of redcare products and services offer both business and domestic users secure and state of the art monitoring networks that constantly monitor every link in the construction chain every second of the day.

A comprehensive range of products are designed for use with a variety of telecommunication services using PSTN, ISDN or GSM with domestic solutions for non-BT lines.

Although these services and the STUs are being regularly changed and amended to suit developments in security, the mainstream services beyond redcare assure that are used in the intruder sector are essentially:

redcare
redcare gsm
redcare home monitoring
redcare mcd
redcare ADSL/broadband
redcare isdn
redcare total

- **redcare.** The essential system with active 2 way monitoring.

 It is a Secured by Design product with a Police Preferred Specification status to the highest Grade 4 of EN50131.

Using high-speed alarm transmission it enables alarm signals to be sent even with the phone in use.

Linking a professionally installed alarm on a customer's premises to an ARC, it is able to detect a line cut and encryption to prevent the substitution of equipment or spoofing (emulation) of the line conditions.

It uses an existing PSTN telephone line to provide a secure communication between the STU and ARC.

- **redcare gsm** combines a secure redcare landline connection with a GSM radio back-up path. In the event of a line cut it enables an alarm system to be monitored by the radio path.

  If the BT landline fails the ARC receives a communication failure. The GSM path will continue to communicate with the ARC sending any subsequent alarms. It is capable of sending a confirmed activation if either the primary landline or the secondary radio path is lost regardless of which is lost first. redcare gsm conforms to governing police policies on confirmed alarms.

> *It is a Secured by Design product with a Police Preferred Specification status to the highest Grade 4 of EN50131.*
>
> *Used for high-risk sites it protects customers' Level 1 police response by automatically delivering alarm signals over the back-up path.*

- **redcare home monitoring** is for residential sites where the back up of an additional monitoring path is required, so it is a communication system which is to a higher level than that which is offered by a digital communicator. It is a simple cost-effective solution for those systems that do not need a communication system to the same level as that of redcare gsm.

> *It is a dual path protection system to Grade 2 of EN50131 and complies with DD 243.*

- **redcare mcd** (modern compatible device) is a plug and play solution for customers who want to use the same BT analogue line for both redcare monitoring and data applications.

  It enables end users to connect redcare alarm monitoring to data applications such as PCs, digital TVs or fax machines, all on the same analogue line.

- **redcare ADSL/broadband.** Customers with an ADSL connection can use the same line to receive the protection and security afforded by a redcare or redcare gsm connection.

  Because these operate at different frequencies and ADSL signals do not interfere with each other, redcare still uses the same continuously monitored low tone 2 minute polling.

- **redcare isdn** is the redcare digital service and is available over ISDN 2e or the Business Highway.

  It allows more information to be delivered quicker and can transmit up to 1000 different alarms or data packets to an ARC over the same connection.

  It is able to pass textual information about a premises' status.

  Compatible with all ISDN protocols it may be used to full capacity for other applications while also providing redcare continuous alarm monitoring.

Figure 6.20   *STU terminal configuration*

It allows the integration of property protection and business applications and provides sequential alarm confirmation, intruder tracking and integration with other security systems such as CCTV.

● **redcare total** is an enhanced maintenance service.

It is available for all redcare connections and provides customers with an accelerated response in the event of a line failure.

It can be specified for critical lines.

We should remember that the STUs are supplied as a PCB that can be installed either by plugging it on to a host panel or hardwiring it in an alarm control panel or in a separate enclosure.

The subscribers' terminal unit STU can be visualized as a communicator that uses the redcare service to pass alarm information to an ARC.

The terminal configuration is essentially as Figure 6.20 and is further explained in Figure 6.21.

Alarm confirmation, as specified by the ACPO policy, can take advantage of back-up signalling such as that offered by redcare gsm. Such a technique may also be referred to as dual path signalling or dualcom (dual communications) as it is a duplicated transmission medium. It is a secure service that can transmit a confirmed alarm if the landline or cellular paths are attacked regardless of the first to be affected.

A STU used to carry out dual path signalling is shown in Figure 6.22. It will be noted that this includes a GSM transceiver with a SIM card carriage and the PSTN line connector block.

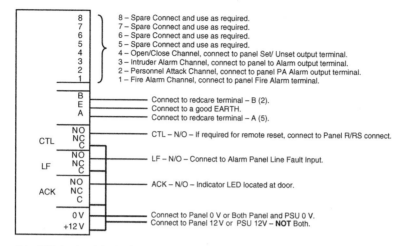

Figure 6.21 *STU terminal configuration/connections*

Figure 6.22 *GSM STU*

Using such a dual signalling technique means that the radio signal can be brought into effect in the event that the normal hardwired line is cut, if the panel fails or if any disruption occurs between the intruder alarm control equipment and the monitoring point. The increase in reliability and

Table 6.4 *STU dual path signalling device pins/channels*

PCB pin No.	Digi channel/ GSM STU	Programmable STU output no.	Default output
1	1	1	Output 1 fire
2	2	2	Output 2 zone PA
3	3	3	Output 3 intruder
4	4	4	Output 4 set
5	5	5	Misoperation
6			
7	Fail to communicate	Not used	–
8	Lo. bat	9 (Lo. bat)	
9	+12 V supply	–	–
10	0 V supply		–
11	Not used		
12	+5 V supply		–
13	6	6	Confirm
14	7	7	Zone omit. Rearm
15	Line fault input	–	
16	Digi reset	8	Digi reset

the rapid advances that have taken place in the cellular networks have led to a far greater use of radio signalling within all communication systems including its use in a supporting role. A consideration is that if the signal strength is weak a better position must be found for those STUs incorporating cellular devices. Higher gain extension antennas and extension cables are available for this purpose.

Before we move on to look at the history of automatic dialling equipment and the digital communicator we can see how dual path signalling is normally applied by referring to Table 6.4. This illustrates the range of allocated PCB pins and channels together with the programmable outputs on a typical plug on STU/dual path signalling device with an industry standard footprint. In this case the inputs are configured to accept +ve triggers (5 V) which must be enabled in the equipment.

Automatic dialling equipment

If we consider the origins of automatic dialling we may well recall the tape dialling machine, which was effectively a tape recorder with a pre-recorded voice on a tape that was connected to an intruder alarm control panel. A signal from the intruder alarm panel started the mechanism on the tape dialler, which was connected to the telephone line. A microchip retained the number to be called and dialled this number as a series of dialling tone electronic pulses. Once the handset at the receiving end was lifted the message was relayed. This device was mechanical in operation, and was the earliest technique employed, although it is now obsolete.

As a progression on this concept we now find that within the domestic sector there is a growing tendency to connect speech diallers to alarm control panels to relay messages to a number of personally programmed telephone numbers when the alarm has been triggered. These are reliable as they are not mechanical but electronic and use a voice recording stored in battery-backed RAM, the personalized message having been recorded by the user. These communicators are mains powered and battery supported. They are easily wired to any intruder alarm panel and can have a number of messages built into them, governed by the signal that they receive from the control panel. These signals may be 12 V, or removed, or be an opening or closing circuit that is voltage free. These devices will continue to dial a selection of preprogrammed numbers until they receive a response, and are plugged into a standard telephone socket. We can now also find control panels that include a speech dialler as an integral part, and these will certainly become more widespread in the domestic sector and may have a role to play in some commercial areas.

A terminal block layout for an intruder alarm control panel with an integral speech dialler that has three trigger inputs, alarm (A), personal attack (P) and fire (F), is shown in Figure 6.23.

The stand-alone or integrated speech dialler has a role to play in some low-security risk applications because it is inexpensive, uses an existing line, and adds no more financial burden to a system. However, for the majority of low- to medium-risk applications the digital communicator remains the equipment of choice.

Digital communicator

The digital communicator is well known as an add-on or plug-in module consisting of a non-volatile memory (NVM), signalling processing circuits and line control circuits. It is programmed by the user, manufacturer or alarm receiving centre to contain the telephone numbers to be called and to include the customer identification numbers.

Channels are selected to give a code at the central station to cover functions such as those below:

Channel	Function
1	Auxiliary, fire
2	Personal attack
3	Burglar, intruder
4	Open/close, set and unset
5	Trouble (setting a system with a zone isolated)

Figure 6.23   *Intruder alarm control panel with integral speech dialler*

Channel   Function
6   Spare (can be used for monitoring of freezers or industrial equipment)
7   Operates whenever the exterior bell rings
8   Undedicated

Before looking at the digital communicator in detail, the telephone line is identified:

5	A	White/blue	BT line
2	B	Blue/white	BT line
3	C	Orange/white	Bell suppression
	E		Electrical earth

These connections are used to connect the digital communicator to the telephone network, which when it goes into alarm transmits a series of pulses to initiate the dialling action to a preprogrammed number. Unlike the tape-type dialler the digital device can sense whether or not the call has been received and ensures that the response force is aware of the activation.

There is no voice communication because the message is sent as a series of encoded pulses that form a type of binary code. This is achieved by a transmitting device that is linked to the non-erasable memory. These pulses are received by a similar device at the receiving station, which then converts them into an alphanumeric print-out or into a video display unit (VDU) format. Due to the absence of any voice communication the problem of poor messages is negated, and because the message is transmitted in a digital form there is no speed restriction as such, unlike the voice message, which must be heard and then understood.

The digital binary code is converted to a tone code by a modem before transmission occurs, and it is then reconverted at the central station by a similar modem.

Devices differ enormously in their capabilities, but do have similar basic sequences which can be summarized:

- Events such as closing and setting the system or detecting intrusion will initiate a trigger signal.
- This trigger signal will be passed from the control panel to the digital communicator.
- The control unit within the communicator must then either connect the signalling equipment to a reserved, outgoing calls only line, or seize a line and inhibit its normal connection to a telephone handset. A dialling pulse generator then sends a dial pulse to a preprogrammed number via the telephone line.
- Having received this signal the receiver prompts the transmitter to give a handshake acknowledgement.
- The dial tone recognition unit awaits confirmation of the handshake. Once received, the control equipment encodes the information to be sent. If this is not received the unit waits for a preset period and carries out a sequence of repeat attempts. This may include repeat calls using the same line and to the same number, or using alternative lines to the receiving central station, or by using alternative numbers or alternative signals. A specified number of attempts are made before a failure report and hang up is generated.
- Providing that the correct handshake is received then the encoded message will be sent via the message transmitter.
- The message is received and acknowledged, and a signal to this effect is returned to the transmitter.
- An off signal is then generated which effectively stops the message repeat signal, which may be decoded and checked before the off signal.

If there is no acknowledgement by the ARC the transmitter will stop, and the unit will repeat the calls to:

- the same number on the same line;
- other emergency numbers;
- the same number but using a different line to the telephone exchange.

An audio or other signal will alert the central station that a message is in the process of being transmitted. The final stages in this process are:

- the message is decoded;
- the message is checked;
- the message is both displayed and printed.

There are many different types of digital communicator available, and selection of the type to use must be based on matching the level of security risk. Multiline, multiple-number devices will always give the highest security and a little more than lower specified models. Digital communicators range from single number, single line, repeat call through to multiple numbers, single line, repeat call; single number, multiple lines, repeat call; and multiple numbers, multiple lines, repeat call. Devices will also be combined with additional techniques to provide extra functions such as point ID extended format reporting or be integrated with modems allowing uploading and downloading to be performed from the central station.

These devices may plug into a professional control panel from the same manufacturer. In these cases the installation is clearly defined. However, stand-alone units designed to operate with all makes of control panel, intruder, fire and industrial, are also available. Those communicators must be suitable for connection to the following types of telephone line:

- Direct exchange lines (PSTN) supporting (DTMF) tone or loop disconnect (pulse) dialling;
- PABX exchanges;
- Relevant branch systems (RBS).

They must also operate with all commonly used receiver equipment. The main specification for a typical stand-alone communicator that transmits data in two (dual) formats (slow 10 baud rate or fast (multitone)) is given below:

Supply
Voltage: 9–15 V DC
Quiescent current: <40 mA
Transmitting current: typically 200 mA

Inputs
8 Channel + low battery (0) + test (9)

Trigger options: normally open contact, normally closed contact, open collector (transistor)
type trigger, normally closed contact removal of +ve.

Received commands
Acknowledge: handshake 1400/2300 Hz
Shutdown: hang up 1400 Hz
Compatible with most receivers

Outputs
Telephone connections: A, B, BC
LED transmission indicator
+12 V switched 100 mA

The digital communicator in which we are interested is supplied either as a circuit board that, where space is available, can be mounted inside a control panel or as a boxed unit complete with a 1 A power supply (Figure 6.24).

If a line divert version is used, the line connections are as in Figure 6.25.

It will be noted that the communicator requires a 12 V DC power supply which must be a BT-approved type and is connected to the terminals 0 V and 12 V+.

Figure 6.24 *Digital communicator: stand-alone format*

Figure 6.25 *Line connectors*

The example communicator described here is an eight-channel version (eight inputs), so there are eight terminals marked as ST1–8 which are used as the connection points for the input trigger channels.

The method of triggering is determined by the information that has been programmed into the communicator PROM.

Latching mode

This means that once the device has started to operate it will always complete the transmission of the input code even if the input is removed from the channel during operation.

Non-latching mode

If the input trigger is removed after the device has started the transmission will be interrupted instantaneously and the receiver will receive a code 9. In this situation the communicator will then abort and make only one attempt to signal this condition to the receiver.

Restore

As the input is removed (reset any time after a successful communication) for any alarmed input, that restoration is signalled by first transmitting the subscriber code, then the alarmed input code and lastly a code (normally 9) showing that the input has been restored.

Triggering

A trigger is defined as rising from 0 V to between 5 and 12 V+. The options are shown in Figure 6.26.

Code selection

Each input channel represents the code that will be transmitted, e.g. ST1 = code 1, ST4 = code 4. There is also a low code priority in that should a second channel be activated during transmission then the input with the highest priority (lowest code number) will be transmitted first. All inputs triggered will be transmitted together during the one 'call' in priority order.

Test

This can be generated by shorting TP–0 V. An automatic test call can be generated at 1–15 hour or 24 hour intervals after the last operation and can be programmed on the PROM.

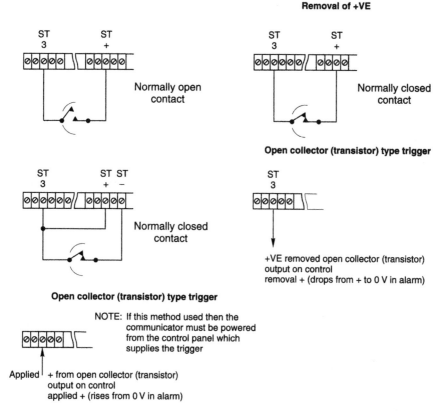

Figure 6.26 *Digital communicator triggers*

Battery low

If this falls below 10.5 V the communicator can activate and report a battery low condition to the central station. This can be achieved by programming any code on the PROM.

CS output

CS and 0 V pins are designed to give a pulsed output on successful transmission of a signal to the central station and is controlled by the program in the PROM.

Prom programming

This is achieved by completing the information on the chart shown in Figure 6.27.

	Primary	Account number	Final account number

PABX digit p a u s e

Telephone numbers

Subscriber code

Secondary

Alternative ☐	Restore code ☐	Low battery code ☐
Dual ☐	Test code ☐	Automatic test call period (1 to 15hrs or 24hrs) ☐ hrs
Tone dialling ☐	Loop dialling ☐	Number of channel eight or sixteen ☐☐

Optional

Start input	1	2	3	4	5	6	7	8	9	10	11	12	13	14	15	16
+ve applied																
+ve remove																
Latching mode																
Non-latching mode																
Restore mode																
Close mode																

Figure 6.27   *PROM programming*

Installation testing

Testing should be performed with the system under full load conditions, using both the normal and standby supplies.

The process of events should be:

(1) Advise the ARC operator.
(2) Conduct the test activation.
(3) Observe that the LED indicator comes on to confirm activation of the communicator.
(4) Listen for the pulses of the dialling relay which should be clearly audible. These should commence some 10 seconds after the activation.
(5) Wait for the LED to go out 5–30 seconds after the dialling has been completed. This is an indication that the message has been received.
(6) Call the central station to verify that it has received the correct code.
(7) Repeat the process for other channels.

Should the call not be received, the communicator will stay on line for 40 seconds, prior to hanging up and redialling. Up to 10 attempts per number will be made before final shutdown.

Fault finding

All communicators will have a form of indication to aid any fault finding that is needed. Two examples are:

Fault: On power up, LED flashes on and off continuously.
Causes: PROM inserted incorrectly; PROM not programmed; PROM not programmed correctly.
Corrective action: Replace PROM having verified programming.

Fault: On power up, LED stays on.
Causes: PROM programmed for +ve removed; fault with programming as normally closed contacts open or vice versa.
Corrective action: Replace PROM, having verified programming.

Under 'Optional extras', mention was made of 24 hour line monitors, bell delay and line monitors and also Selectaline scan monitor. These are detailed next.

24 hour line monitor

This is recommended for ex-directory incoming calls barred telephone lines. It is a circuit board intended to plug onto the communicator main board and continually checks the line voltage every 5 seconds even when the communicator is transmitting. If no voltage is present (under 3.5 V across the line) for approximately 20 seconds, the relay contact will change over and the fail outputs will operate.

Terminals will be provided on the board for:

- the SPCO relay contacts (three terminals);
- the fail output – floating going to 0 V when triggered (one terminal);
- a 12 V positive for reference to the fail terminal (one terminal).

The operation of outputs will be controlled by the configuration of jumpers on the board, and will give features such as:

- *LF*. Relay follows line condition. The fail output will override the line condition and await a manual reset.
- *N*. Line fail will cause the relay to latch until reset, the fail output operating as LF.
- *MAN*. To reset, the jumper is removed and then replaced.
- *AUTO*. Reset occurs when the communicator triggers.
- *L/C FAIL and F. LATCH*. The relay will operate on line fail and communications fail, the fail output will also operate on line or communications fail but will latch until manual reset.
- *L/C FAIL and RESET*. Both the relay and the fail outputs will follow the line condition.

- *No jumper and F. LATCH.* The relay will follow the line condition, the fail output will follow communication fail but will latch until reset.
- *No jumper and RESET.* The relay will follow the line condition, there will be no fail output on communication fail.

Bell delay and line monitor

These are designed to be used with most alarm control panels using either an ex-directory incoming barred line or an ordinary domestic telephone line. The line monitor will operate successfully with personal alarm, open/close and auxiliary alarm communicator inputs, as well as the normal intruder signal. This is achieved by signalling the bell delay separately by connecting the intruder input with the delay bell (DB) input start. This allows other signals to be transmitted without bell delay operation. When operated, the bell delay can be simply reset by removal of the input, simplifying any system reset operation. Bell operation with a link selecting 'Norm' will be suppressed until the preset delay has expired. This mode of operation is satisfactory with simple panels but undesirable if more facilities are used requiring instant bells.

Inverted bell operation with the link in the 'Inv' position is generally used with more sophisticated panels incorporating bell hold-off and bell timer functions. This has the advantage of not suppressing the bell if required on other alarm signals.

The inverted position can be used to provide a bell timer function when used with simple panels. This mode of operation will always give instant bells. For intruder alarms and managerial signals, such as test, automatic test and open/close signals will not operate the bell.

In most applications the DB input will be connected to the intruder input, so when the communicator is activated the bell delay will start running. The 24 hour line monitor is brought out on separate contacts for wiring into the 24 hour panel circuit, or to activate an annunciator, if required.

The line monitor is line following, and will automatically reset when the telephone line is restored. The purpose of the board is to provide 24 hour line monitoring and bell delay with line analysis. The 24 hour monitor is intended to register a fault within some 20 seconds when:

- telephone wires A and B are cut together or are cut separately;
- telephone wires A and B are shorted together.

The bell delay is capable of being triggered by either a positive applied DB+ or removed DB− input, with an automatic reset on the restoral of the inputs to the quiescent state. The bell timer can be programmed for increments of up to 20 minute intervals. The bell delay will automatically cut off when one of the following events occur:

- incoming ringing while communicating;
- off hook while communicating;
- failure to communicate within 3 minutes.

We can summarize the above as the bell delay relay being selected to operate in two modes:

- *Norm.* Contacts change over after a preset delay when triggered.
- *Inv.* Contacts change over immediately, and after a preset delay return to their quiescent position.

Selectaline scan monitor

This is intended for use with the digital communicator main board, and is used for more secure applications being connected to two PTT telephone lines, which are best brought to the signalling equipment by different routes to reduce any chance of tamper and sabotage. In addition to having the features of the previous units, this monitor will also check for busy telephone lines. If the incoming signal (current sensing) is present or the line is already engaged (handset off hook) (voltage below 32 V) for 4 seconds when the device operates, both changeover contacts will operate.

In addition to having bell timing functions and reset of line fault restore, the Selectaline scan monitor will also choose a serviceable line from two alternatives. Normally, an ex-directory incoming calls-barred PTT line is connected to terminals identified as A1, B1 and C1, and an ordinary service PTT line with perhaps a handset connected for daily use is wired to terminals identified as A2, B2 and C2. When an activation occurs, the communicator will try to signal on the A1/B1/C1 line first. If, however, a fault occurs on that line it will select A2/B2/C2 and attempt to communicate. If this is unsuccessful it will again revert to the first line, and continue to communicate in this way between the two lines until either:

- success in reporting is achieved (either single, alternate or dual);
- the transmitter completes its programme of attempts, which will be of the order of 10 for single telephone number, 19 for two telephone numbers or as generally programmed to instructions.

Digital communicators vary enormously in their capabilities, although many of the functions we have attributed here to 'add-on' devices will in some instances be employed on the main board. However, there are a number of potential problems that exist with digital communicators and should be understood. One of the considerations must be the reliability of the PSTN, and the other main problem is special information tone clash. There are of course many ways to transmit an alarm signal to a central station although the public telephone company voice grade line is still the most common. One major problem with these lines is that they are not monitored for disconnection, and parallel alternative lines should be employed whenever possible. We do of course recognize the advantages of redcare, which both checks and monitors the system at all times, and if either the alarm is activated or the line cut, an appropriate message is sent to the central station.

When redcare is not used, a method of improving the security of a digital dialler is to use a digital communicator looped system. With these there are effectively two outputs – one links the direct normal line to the central station and the other links a neighbouring communicator. This neighbouring communicator has a similar but separate line to the same central station.

Should the normal line fail, the communicator will pass calls through all of the communicators to which it is connected until it establishes a line over which to transmit the message. Although not as secure as a multiline multinumber communicator on the same premises, it can always be seen as an alternative to a direct dedicated monitored private line. The problem of special information tone clash can occur when a digital communicator attempts to ring a number using the PSTN. When the line is busy it will stop and try to ring again. Special information tones (SITs) introduced by telephone companies to precede messages and advise that there is congestion are unfortunately similar to those used by central stations to indicate that a message has been received, and can cause confusion. Fast and superfast dual-frequency diallers do not suffer with this problem in the same way as slow format types, and are advocated. At this stage we can summarize the requirements of a digital communicator:

- compliance with BS 6320, *Specification for Modems for Connections to Public Switched Telephone Networks*;
- battery capacity for five separate alarm call sequences (or 20 if the supply to the recharger cannot be guaranteed);
- a battery recharge within 24 hours;
- a 1 second initiation of call sequences from alarm recognition;
- the communication open-channel signal to be received within 1 minute.

If this does not happen, the device should hang up and restart the sequence by means of a different line/number.

Once the communication line is established, up to 10 messages may be sent in an attempt to obtain an acknowledgement signal. If after 10 attempts this acknowledgement is not received, the line is to be released and the whole sequence repeated for a maximum of three times. If after the third attempt the message acknowledgement signal is still not received (and the system has no deliberately operated devices) a local alarm should ring within 10 seconds.

If a fault, off hook, phone-in, or other block in the monitored line is detected (and the system has no deliberately operated devices), then the system will sound a local alarm within 30 seconds.

With the redcare installation a subscriber terminal unit (STU) is the interface between the alarm system and the telephone network and is used to transfer the panel status information to the central station. The wiring technique is not unlike that used for the digital communicator, with the control panel outputs connected to the related channel inputs on the STU as normal. The power to the STU is either derived from a BT-approved power supply or from the control panel as applicable.

Discussion points

We should recognize the benefits of underground lines, separate lines for an alarm system or an incoming calls-barred line or ex-directory number. The primary requirement for line fault monitoring (LFM) is for the overriding of the bell delay as applicable. If there is no bell delay then the necessity to monitor the line is reduced. The secondary benefit of LFM is to advise the alarm system user that the line is out of commission so that the telecommunications company can respond. The requirement to monitor off hook to tie up the alarm line whilst an intrusion is effected is not so important with the advent of digital exchanges, as these automatically monitor the active lines and if there is no activity on a line for a period not exceeding some 10 minutes then the line is reinstated.

The important consideration for the installer is to ensure that the client is aware of the signalling medium and what policy to follow in the event that the signalling path could be lost. In so far as liabilities are concerned, the installer's exclusion of liability must refer to the supply, maintenance and repair of the remote signalling media.

6.7 Radio signalling

Paknet originally launched an alarm signalling service using radio frequency to the security industry in 1990. It was timed to meet the challenges that faced the industry early in the decade. It was felt that a need existed to revolutionize existing alarm signalling services, then dominated by digital communicators on unmonitored lines.

The alarm system can only ever be as secure as the communications link to the alarm monitoring station, and it was believed that using radio signalling with its immunity to line-cutting attacks could be a major step to secure communication. This was achieved by high-speed digital radio transmission through a national network of radio base stations.

Radio as a signalling medium can be:

- *Fast.* Transmission in under 5 seconds. Less time is allowed for intruders to damage or interfere with the alarm panel before transmission of the alarm signal.
- *Secure.* Immune to line-cutting attacks. Invisible and physically inaccessible from outside of the protected premises.

- *Resilient*. Duplicated network structure. Base stations have at least two radio channels, therefore the system has at least two separate paths from each location to the central station. In some cases there may be four or more links where the radio pad can obtain service from more than one base station.
- *Immune*. No telephone renumbering. No reprogramming needed in the event of changes in national dialling numbers.
- *Flexible*. Quick and easy to install. Can be adopted with new or existing alarms. Control panels are available with the radio pad built into the panel.
- *Digital*. Designed for short bursts of data – ideal for the transmission of very fast alarm messages.

Method of operation

Installation requires the fitting of a special interface device and a small antenna. The interface device links directly to the alarm panel and contains the radio pad which sends and receives the alarm information over the radio data network. On alarm activation the interface instructs the radio pad to send a short message via the nearest radio base station through to the central station. The central station will return an acknowledgement code and also contact the emergency services.

Since the system is not susceptible to line cuts, as the communication path is by radio and not wire, there is no need to poll the links constantly. Nevertheless, the quality and performance is checked remotely every hour. A service availability indicator is always displayed on the alarm panel as an assurance.

Statistics have always shown that for commercial risks the telephone line is liable to be attacked and cut. A method of dual-signalling sequential confirmation using radio backed up by a digital dialler can accept this risk and balance it by giving a secure combination. Radio added to an existing digital communicator system can monitor the telephone line for signal failure or line cuts while the digital communicator monitors the radio link. Each service essentially monitors the other. The radio interface informs the central station of the line being cut. Considerable development focus has been placed on the radio pads, which are available as original equipment manufactured modules for easy integration into existing control panels. The polling service to them on an hourly basis is totally automatic and provides the alarm receiving centre with key management data, and the installation is provided with even a specific check method to ensure that the aerial is in the optimum position.

In summary, high-security transmission can be achieved with radio used in conjunction with a digital communicator offering dual-signalling communications. It competes with redcare GSM.

6.8   Optical fibres

Fibre optic transmission systems are multifunctional. A single cable running around a building can transmit not only alarm signals but also guard intercom signals, CCTV signals, audio signals and log data from patrol points, lighting system relays and access control information.

Fibre optic systems are inherently safe to use, and have a very high immunity to most harsh environmental conditions. Fibre optic cables, however, are more expensive compared to other recognized cable types but do have good qualities with regard to interference resistance, safety and also transmission quality and speed. Fibre optics comprise cores or filaments of extruded glass in a continuous length. This glass is essentially pure with high transmissivity to light along its length. This property of linear light transmission is used to send data in an optical form. Thus, energy in the visible part of the electromagnetic spectrum is used to transmit data instead of energies of microwave frequencies. Light travels at the same speed as microwaves (300 m/s), and is both easily generated and controlled.

Optoelectronic technology can be defined as a technology encompassing devices that function as electrical-to-optical or optical-to-electrical transducers capable of responding to light. In practice, a fibre optic transmission system (Figure 6.28) will comprise a transmitter, a fibre optic cable, a photodetector/receiver and perhaps a reception amplifier.

In operation the transmitter will accept an input voltage signal from a particular system, and it will then transform this signal into a modulated beam of light. This light is then focused within the fibre optic cable and transmitted at the speed of light to the photodetector/receiver, which will in turn convert this light input signal into an electrical output. The photodetector/receiver comprises a highly light-sensitive chemical receptor

An optoelectronic unit comprising a driver and source which can change electrical signals into optical signals.

A cylindrical core of flexible transparent material surrounded by tubular cladding of material with a lower refractive index. Light enters at one end of the optical fibre and emerges at the other end, after propagation by internal reflection along the core.

A detector which changes a light input into an electrical output signal.

Figure 6.28   *Fibre optic transmission system*

which can convert the coded light energy into an output signal by means of an amplifier.

The frequency of transmission and the bandwidth depend on the application, but are in the range of 10 Hz to 10 mHz for CCTV systems and 10–20 Hz for audio transmission.

Bandwidth is the difference between the upper and lower frequencies which can be sent along a communication channel. When used in radio engineering, bandwidth is quoted in cycles per second or hertz, but the word bandwidth is now also used to specify digital data transmission channels, and hertz then refers to bits per second.

Uses

Optical fibres are particularly useful as links between buildings as they avoid the risk of lightning damage and electromagnetic interference (EMI) or the noise that could be generated in electrical conductors by stray electromagnetic fields. There is also no concern with electrical isolation by using fibre optic links. These links can transmit the following simultaneously over the same cable either in the uni- or bidirectional mode:

- audio signals;
- video signals;
- alarm signals (contact closure data);
- pure data;
- telephone voice grade.

Within the intruder alarm industry the predominant security use of fibre optic links is in CCTV, particularly in harsh environments where the cables can be buried as multiplexed runs to central stations.

Fibre optic transmission systems can provide cost-effective high-quality generation of video signals over multimode fibre for fixed CCTV applications such as those found in security and surveillance sectors, and these can then be integrated with intruder alarms (Figure 6.29).

The transmitter is a small unit that can plug into the BNC port at the rear of the surveillance camera. A standard photodetector/receiver will have a number of separate video channels, allowing operation over some 3 km.

Transmission loss

The attenuation or loss measure of optical power along an optical fibre is measured in decibels per kilometre, and quoted at a specified wave-length of light. The calculation of total transmission loss (TTL) in a system is of huge importance and takes into account linear attenuation of the cable, the attenuation of all the joints or splices and also the coupling loss at the cable ends plus an aging loss allowance.

Figure 6.29   *Fixed CCTV and intruder alarm system*

In any system there is a maximum allowable loss (MAL), and the TTL must stay within this. Light is a very concentrated energy form, and photosensitive sensors are extremely sensitive, so an optical fibre system can still operate with a 90 per cent loss of transmitted light energy. Nevertheless, the number of connections should be kept to a minimum and the most cost-effective joint method used for any particular application.

There are various connector forms, ranging from purely mechanical with no fusion, through to fusion by a heat process, or jointing by a hot melt, or heat-cured epoxy-terminated, process. The selection of these is also governed by the cable selection type.

Advantages and disadvantages of fibre optic links

Advantages

- All signal transmissions possible.
- Little attenuation or loss throughout the cable itself.
- Up to 5 km cabling without amplification.
- No interference from lightning, Radio Frequency Interference (RFI), EMI or radiated energy.
- No explosion risk or fire or sparking possible.
- No corrosion.
- Can withstand harsh environments and can be buried.
- Huge data and simultaneous transmission capability.
- Small diameter for capacity.
- Strong when under tension.
- Useful in potentially explosive atmospheres.

Disadvantages

- Cost.
- Limits on bending.

- Can be crushed and damaged (115–125 kg/cm are typical limits).
- Operating temperature ($-10°C$ for indoor and $-40°C$ for outdoor cable).
- Care needs to be exercised in terminating and jointing (inherent cost).
- Not compatible with nuclear radiation environments.

Connectors and installing

In the past the fitting of connectors involved the use of epoxy resins and ovens to cure the resins followed by many hours of polishing to ensure a perfect finish.

With new-generation equipment, excellent terminations can be made with crimp-type connectors similar to those found on co-axial cables. This standard crimping process is followed by light polishing. Kits to do this work can be hired to alleviate any early outlay.

Setting up

Fibre optic systems, unlike copper wire-based systems, attenuate the signals equally irrespective of frequency. Therefore, rather than having several controls to balance the signal after transmission, all that is needed is simple amplification. This is carried out by a process called automatic gain control (AGC). This is performed in the receiver units, and the same electronic circuit ensures that the signal level is adjusted throughout the life of the equipment. This makes the unit essentially free from adjustment.

The intruder alarm industry at present may well be in its infancy with respect to optical fibres, but their use will surely become more established in the future. The increasing use of optical fibres in the transmission link will inevitably change the way security installers work.

At the present point in time the use of optical fibres may well be seen as a specialist subject, but the importance of their applications and uses should certainly still be understood.

6.9   Alarm confirmation technology

The electronic security industry is facing significant changes in the way in which the systems operate. These changes are due to the various police forces' response policies for alarms from electronic systems in that the security industry has been forced to reconsider its false alarm management methods to the extent that alarm signals are to be confirmed. Risk analysis also assumes increased importance for installation companies working to the criteria of the new policy. Systems must become more intelligent to ensure that remote intruder alarm systems continue to obtain a police response in the future.

In light of the need to move ahead DD 243 was introduced on a provisional basis to enable the information and working experience gained from it to be established in a practical way. It is the code of practice giving guidance and advice on the design, installation and configuration of intruder alarm systems signalling confirmed activations to an ARC. DD 243 only involves new systems or systems that are being reinstated after response has been withdrawn as existing systems are exempt. This document is not actually exclusive to systems requiring a police response as any site can benefit from the eradication of false alarms. It is also important to recognize that the ACPO Security Systems Policy (originally termed Intruder Alarm Policy) does not seek confirmed alarm systems but it seeks confirmed activations. To this end there are three methods of confirmation:

- *Audio*. Following an alarm activation the ARC operator listens in to the protected premises to ascertain whether or not the alarm is genuine.
- *Visual*. The ARC operator analyses recorded images from the protected site to ascertain whether or not the alarm is genuine.
- *Sequential*. This is a technique by which a second activation must come from an independent detector to the activation that was received from the first device. This second activation is to come within a defined time window so as to indicate that an intruder is moving through the premises but is not on an entry or access route.

Confirmation also extends to signalling in that a confirmed alarm may be generated in a number of ways providing that dual path signalling is installed. This is by the means of:

- transmission fault followed by an alarm signal
- alarm signal followed by a transmission fault
- two transmission faults.

Audio confirmation

Audio confirmation can be carried out by either audio listening devices (ALDs) or audio monitoring devices (AMDs). The former devices allow the ARC to listen in to the site having initially been triggered by an intruder detector. AMDs are activated when the ambient sound exceeds a threshold so they are essentially a single device providing the dual role of monitoring and listening. The considerations are that:

- audio cover matches the detection coverage
- the audio confirmation does not commence until the ARC has been signalled
- when an alarm is received at the ARC the listening periods are as a minimum

−60 s for intruder detection

−30 s for audio detection

- for audio confirmation of intruder detection there is to be 10 s pre-recorded stored audio before any activation and at least 15 s stored after the activation
- for audio confirmation of audio detection there is to be 1 s of stored audio before an activation and 15 s stored audio after an activation
- with an automatically established link the device is to store an equivalent period of audio to the time taken to establish the link
- audio devices are not to be located in areas of high ambient noise
- part set modes are to isolate both the detection devices and the audio devices in the unset area.

There is a growing range of systems available for the installer to open up an appropriate voice channel at the protected premises and to detect sound in the area where the alarm has been generated so as to enable the ARC to listen in. Success is very much related to the number of microphones deployed and to the speed in opening the channel. Remember that there is no need to install microphones in areas that have no detector.

Video confirmation

This is carried out by imaging devices activated by a linked detector or by video motion detection (VMD) or a mixture of both techniques. With this technique the video cover must match the detection coverage and the video signals are not to commence until the alarm has been generated to the ARC. There is to be a minimum of 3 images following the activation – 1 image at the time of the activation followed by 2 images after the activation and within 5 s.

This technique can be centred around technologies such as fastscan that allow picture information to be relayed to the monitoring point using a custom modem as an interface with the transmission medium. ISDN lines are an effective medium for picture transmission and because of reductions in the costs of these links we shall continue to see an increase in the remote monitoring of sites. Figures 6.30 and 6.31 illustrate how the system can be applied to a system with purpose-designed unique camera detectors or to any CCTV network.

Sequential confirmation

This is an event where two separate signals are reported from independent detectors within a given period of time. It means that the sensing devices are connected to separate zones on the control equipment by separate outputs and connections. Although two distinct zones must be employed, the detectors can be in the same room, but if they are of the same technology

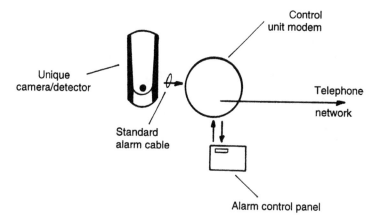

Figure 6.30 *Alarm confirmation configuration*

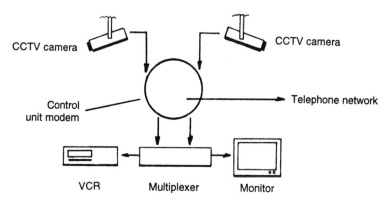

Figure 6.31 *Remote monitoring*

such as two PIRs they are not to have overlapping areas of cover. This is illustrated at Figure 6.32.

Therefore an alarm signal will generate a confirmed signal when the two independent detectors operate within the programmed confirmation time. This shall be 30 to 60 minutes. It means that the engineer can set the programme of the control equipment so that following the signal from the first detector the second detector must then respond within a time of up to a maximum of 60 minutes later. In all cases there is a need to signal the ARC as to the system status (i.e. system set or unset) and to meet the following criteria with respect to detection devices:

- Different technologies may overlap. In the case of two movement detectors at least one should be a dual-technology detector (e.g. PIR/microwave).

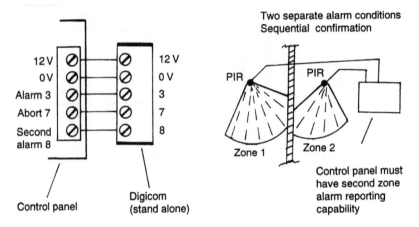

Figure 6.32   *Alarm abort and sequential confirmation*

- Same technologies are not to overlap but two dual-technology movement detectors having the same technologies may be classed as detectors of differing technologies so may overlap.
- Detectors are not to be triggered by the same event, such as two door contacts on the same door being wired on separate zones.

For sequential confirmation, an alarm condition from the first sensor should generate an unconfirmed activation. If there is a subsequent alarm within the programmed confirmation time span from a separate detector a sequentially confirmed activation can be generated. As a typical example this may be a door contact activating followed by a movement detector. If the second (confirmed) detector does not activate within the programmed confirmation period the following is to apply.

- Reinstatement of the first detector if possible
- If not isolate and signal isolation to the ARC
- If a perimeter detector the ARC to carry out keyholder action
- If the entry door is isolated an alternative means to initiate the entry timer is required (a secondary means of entry could be through an 'Access' route that could effectively become 'Entry' if isolation of normal entry has been applied).

In order to ensure that confirmed activations cannot be caused by user error there are specific setting procedures which must be met and these are detailed in the part of Section 1.5 dealing with false alarm prevention. Additionally, when designing for sequential confirmation it is necessary that the equipment can signal a secondary signal in the case of user error and in order to abort and terminate the operation. Figure 6.28 shows a

Table 6.5 *Typical point ID transmissions*

Site ID 2250	Set	User 25
Site ID 2250	Alarm	Circuit 07
Site ID 2250	Alarm	Circuit 02
Site ID 2250	Unset	User 11

stand-alone digital communicator with alarm abort and sequential confirmation. We can describe the operation as follows:

● *Alarm abort.* When the alarm is triggered, the system transmits a channel 3 (alarm) to the alarm receiving centre. If the system is unset within the 'abort delay' period the abort output is triggered, and channel 7 (abort) is transmitted to the ARC.
● *Sequential confirmation.* When the first alarm is triggered, the system transmits channel 3 (alarm) to the ARC. If a different zone is then activated, the system will transmit channel 8 (second alarm) to give a confirmed alarm. If there is no second signal it is treated as a normal alarm.

Point ID extend format reporting is an enhanced form of sequential confirmation via a digi-modem and provides more detailed information on the alarm. It creates a communication path and allows an operator to view the status of each individual alarm zone. Table 6.5 shows typical transmissions. These appear as text at the ARC. They cover circuit identities, user identification and system status.

The *order of events* using such transmissions can be:

(1) user 25 sets the alarm and the influence is relayed to the ARC
(2) the alarm is activated and the system transmits the site, activation type and circuit that caused the alarm to the monitoring station
(3) the ARC recognizes this initial signal
(4) circuit 2 activates and the police are informed
(5) user 11 unsets the system and no further activations occur.

It follows that sequential confirmation is a system feature that requires control equipment that has second zone alarm reporting capabilities. However, most existing systems can be upgraded to signal confirmed alarms by fitting new control equipment complying with DD 243: 2004 to carry out sequential confirmation whilst retaining the original detectors and wiring.

If we refer to Figure 6.33 it is possible to see the basic operation of a sequentially confirmed system and how the communication outputs would be programmed on a typical STU. It is to be noted that different channels may apply to the equipment offered by alternative and competing manufacturers although channel 3 is traditionally intruder.

Basic operation

Intruder (03)	At first alarm
Confirmation (08)	At second valid alarm
Zone omit at rearm (17)	In the event of a zone being omitted at 'Reinstatement' of the system at the end of confirmation time
Set/unset (04) OR	When system sets/unsets
Misoperation (07)	When system manually unsets following an alarm

Programming communication outputs

Pin	Loom colour	Prog. O/p	Function
1	Brown	*	Line fault input
2	Red	*	RedCARE reset or fail to comm. input
3	Orange		12 V supply
4	Yellow	07	Misoperation (abort)
5	Green	17	Zone omit at rearm
6	Blue	08	Confirmed alarm
7	Mauve	03	Intruder
8	Grey	04	Set all
9	White		0 V supply
10	Black	02	PA

Figure 6.33   *Basic operation sequential confirmation to ARC and comm. outputs*

6.10   Signalling and active goods protection

To this stage we have considered a range of types of signalling covering audible, visual and remote techniques. We also know that once an intruder alarm system has been activated it can carry out a range of other functions. One of the progressive techniques is that of fogging systems which is referred to as active goods protection.

Tags on goods, intruder alarms and CCTV are powerful deterrents but fog delivery systems have a huge supporting role to play in asset protection to counter ram raiding.

Shop owners need to comply with strict insurance requirements with traditional deterrents such as reinforced glass, ram raid bollards together with intruder alarms and CCTV linked to an ARC. However, once the intruders gain access to a premises a window of opportunity exists until the authorities arrive. A long response time can have serious implications to the loss of merchandise to burglaries such as ram raiding because these intrusions may be completed within a few minutes.

The answer to this problem is to close the window of opportunity and to protect goods of value by using active protection until the authorities arrive. We also know that under the banner of active security we have rising security screens and rising bollards that are automatically raised after closing time in commercial and retail premises.

Smoke generators have been employed in the entertainments sector for many years and it was this essential technique that led to the generating of a blanket of impenetrable fog that is now used in the security industry.

In practice, a vapour is produced by machines from a safe and harmless mixture of water and glycol. When the alarm is activated and the fog required, the mixture is quickly heated to a high temperature by passing it through a heat exchanger which is permanently charged at all times. The formula is emitted into the environment under high pressure whereupon it immediately vaporizes into dense fog.

This rapid fogging leaves the intruder shocked and disoriented and unable to see the room contents. By a link between the fogging system and the alarm the dense fog is triggered to activate at the exact time the signalling is brought into effect.

The fog can stay suspended for a period in the order of an hour, can refill the room if necessary and be used safely in any environment and where any form of goods or foodstuffs are being stored.

In general terms a flume of fog can be propelled a distance of six metres in two seconds and fill a room of some five metres by ten metres and of a standard height of two and a half metres in under five seconds.

It becomes clear with all systems that carry out active functions that there can be no room for false alarms, and equally the detection devices in use must have high levels of capture ability. Confirmation techniques can be used to prove that intruders are moving into a protected area.

The signalling can also be as a response to a PA hold up output or in the case of rising bollards to follow the timed setting of a security system.

In all other respects the actual cabling and programming of the active security is not complicated.

7 Intruder alarm wiring systems

Regulations cover the segregation of cables of different categories and the susceptibility of electromagnetic interference.

Good practice is required to ensure the neatness of the finished system and mechanical protection of vulnerable wiring. This chapter provides knowledge of the techniques involved in installation practices.

This chapter familiarizes the student with the constructional features of the wiring conductors that are employed in the security industry, and ensures that their applications are understood. The types of terminations and connection methods that are generally encountered are also discussed.

Cables will be found alongside flexible cords in a number of different guises. These can be PVC (polyvinyl chloride) insulated or sheathed and may also be fire resistant. The terminations and joints will be soldered or mechanically formed, and so will have differing mechanical and electrical properties.

In addition to citing the standards governing the joining of wiring, some study must also be made of how these standards relate to the segregation and support of wiring. The installation of wiring systems must therefore cover the methods of fixing to various surfaces and the reasons for any particular selection, including the use of conduit and trunking. Allied to these needs is a requirement to be aware of the methods that can be employed to fix alarm equipment to the various surfaces that will be encountered, and an important area in this regard is the mounting of external audible intruder alarm signalling devices.

Also within this chapter we investigate the use, care and maintenance of ladders and stepladders.

7.1   Survey of modern wiring methods

We must first consider how the chief methods of electrical wiring systems are applied throughout the electrical industry before we can look directly at the ways the wiring can be employed in the intruder alarm sector.

A general survey of the chief methods is appropriate to aid the installation engineer in helping him or her determine the particular wiring technique most suited to any given application. Reliability and ease of installation both come into the equation, together with the cost. In the intruder alarm areas we are accustomed to particular cables being utilized, but there are many different ways that these can be protected. All systems are to comply with the environmental conditions that exist at the supervised premises. These conditions include potential causes of mechanical damage as well as weather and heat. Problems can also occur as a result of dampness, corrosion, oil and electrical interference or because of an adverse industrial atmosphere. Indeed, this standard imposes stringent requirements on the installer, and it becomes essential that room data sheets include records pertaining to the environment. Persons commissioned to install intruder alarm wiring may not of necessity be qualified electricians but nevertheless the standard of work must be to the same high level. With this in mind we shall consider how wiring is installed and protected in the electrical engineering sector.

Steel conduit systems

With this method steel tubes are fixed to the walls and building structure, and cables are drawn into them at a later stage. The cables are usually PVC insulated and sheathed or non-sheathed, and, for mains wiring techniques, may be single core. Although the tubes can be fixed to the surface of the building, the full advantage of this method is realized in a building which is in the course of erection as the conduit may then be fixed to unplastered walls or chased into brickwork before being encased in plaster. The cables should not be drawn into the conduit until the plaster has had some time to dry, to prevent the ingress of any moisture into the tubes from the wet plaster.

Types and grades

Conduit is supplied in standard lengths of 3.75 m in diameters of 20 or 25 mm conforming to BS 4568. Steel conduit is black enamelled or galvanized, and threaded at both ends. For aggressive environments high-grade 316 stainless steel conduit is available; this type of conduit is able to cope with chloride-laden conditions, and is therefore ideal for applications in the food, brewing, dairy, pharmaceutical and chemical industries. As stainless-steel conduit also has an aesthetically pleasing finish it can be utilized for purely architectural reasons.

Non-metallic conduit systems

Although recognizing that steel conduit provides excellent protection of all cables to mechanical damage and also gives earth continuity, there are certain disadvantages to its use. These largely relate to condensation, rusting and corrosion of the material. In the long term this can lead to decreased protection and loss of efficient electrical continuity. It is therefore always worth considering non-metallic conduits, which are available in either rigid or non-rigid forms. They are made of PVC capable of providing high-impact resistance, and are usually available in 4 m lengths with the same diameters as steel conduit (20 or 25 mm).

PVC-based conduits have the following advantages:

- High resistance to corrosion by water, acids, alkalis and oxidizing agents. These materials are also unaffected by the chemical components in concrete and plaster.
- Dimensionally stable.
- Do not deteriorate significantly with age or external exposure.
- Non-flammable.
- Not susceptible to water condensation.
- Excellent electrical properties, with an electrical breakdown voltage of the order of 12–20 kV/mm.

Rigid conduit

The most commonly used type of rigid conduit is 'unplasticized' PVC conforming to BS 4607. It has plain bored ends and two wall thicknesses, either heavy (standard) or light gauge, and is generally coloured black or white.

In practice, heavy gauge conduit is installed on surface installations where mechanical damage is a distinct risk. The normal method of joining and applying fittings is by the use of push fitting. The push fit conduit entry ensures a tight, reliable fit, and when used in conjunction with PVC adhesive a strong permanent joint can be attained. For protection in damp conditions, solid rubber gaskets can be employed.

Flexible plastic conduit

This is available in long coiled lengths usually of 25 m, and is used for sunk or concealed wiring in cases where its appearance is not of importance. It has great flexibility and enables awkward bends to be negotiated, and can be threaded through holes easily. The conduit can be applied over irregularities in wall surfaces without difficulty, and can withstand the stresses imposed upon it when floors are awaiting screeds. Flexible conduits are normally of 20 mm diameter.

Trunking

We have said that the erection of conduit should be done before running in the cables; however, there is an alternative form of cable protection available which permits the cables to be installed even prior to the protection being fitted. This is PVC trunking, and it is widely used by the intruder alarm engineer. It can even be applied at the very last stage of an installation or to an existing application. This is a form of channel manufactured from high-impact PVC and featuring a locking or double-locking lid that is pushed into position and held within its longitudinal channel ridges. Trunking is normally supplied in a white finish, has a high aesthetic appearance, and is ideal for most types of electrical installations from security to fire and lighting and power systems. It is manufactured to comply with BS 4678 and can either be held in position by fixing its back channel section in place with plugs and screws or using a self-adhesive foam strip, if the surface to which it is to be applied is free from dust and grease. Outlet boxes to complete the installation are also available to ensure an adequate level of mechanical protection as required by the IEE Wiring Regulations. Trunking is generally available in standard lengths of 3 m, and is easily cut to size. There is also a range of sizes available to suit the cables that are to be accommodated, the following being the most commonly encountered:

Width (mm)	Depth (mm)
16	12.5
16	16
20	10
25	12.5
25	16
38	16
38	25
38	38

As with conduit, a range of accessories can be purchased to make up the installation. These range from couplings that are used to join the lengths of trunking to box adaptors, 'T' pieces, flat angles and internal and external angles to negotiate bends and corners or to make up branches. Blank ends are used to terminate runs and to enclose the end of the trunking for both strength and aesthetic purposes.

A range of PVC mini-trunking can also be sourced, which is supplied in a dispenser box as a flat coil some 15 m in length. It is easily cut to the desired length, reducing waste and the necessity for excess joints and couplers. It is fixed to the required surface in a flat form, and the sides may then be folded up and the lip clipped into place. Once again, accessories are available to extend its use, including outlet boxes.

Although the intruder alarm engineer will rarely use galvanized trunking and fittings, they are occasionally encountered in industrial environments, where they are used to protect mains supplies. These are manufactured from precoated galvanized steel with the lid fastened by integral fixing bars that engage the trunking body when the captive lid screws are rotated through 90°. Another product is the standard flange tray, which is designed to carry heavy cable installations. Alarm cables must never be run in close proximity to either galvanized trunking or flange trays when carrying mains supplies.

There is one further form of support that can be used for alarm or signal cabling, known as dado or bench-type trunking. This is effectively installed as skirting and is an effective means of running cables at a low or bench height. This type of protection tends to be manufactured either in PVC or sheet steel with a white epoxy paint finish. Again, accessories to complement the trunking are available.

Aluminium tubing

A further form of cable protection that is at times used for alarm wiring is that of tubing using aluminium tube of 12.5 mm diameter. This method is in fact a long-established means of protecting taut wiring used as a detection device for openings such as windows. The tubing can, however, also be found protecting individual cables in some installations where trunking is not practical, such as in external applications where the superior appearance of the tubing and total enclosure of the wiring are wanted. Aluminium tubing does not corrode and therefore has a role to play in harsh environments. It is purchased in 3 m lengths and is easily cut to size. It is joined by clamp couplings that feature screws and bolts that fix the coupling ends over the tubes being joined. Elbows to negotiate bends and changes of direction are also applied by a clamping technique, and saddle clamps are used to fix the tubes to the building structure.

PVC channelling/capping

PVC-sheathed cables running along walls may be buried directly in the plaster but they are better protected by placing metal or PVC channelling over them. This gives a level of resistance to nails being driven into the wall, and in the case of conduit permits the cables to be withdrawn at a later stage if so desired. The channelling we have mentioned is sometimes called capping, and is generally used for the protection of electrical installation cables when surface wired to brick/block work prior to plastering or rendering. Supplied usually in PVC material in a variety of lengths, it is so flexible that it can be carried in reels; it is also shatter-proof, so it will not crack when nailed in position. It takes up the contours of the wall and is easily cut to length using cable cutters or shears. It is manufactured from insulated self-extinguishing material having low smoke, low toxicity and low acid emission properties. Various widths are available to accommodate different volumes of cables. Once the cables have been installed, the capping is simply placed in position over them and nailed into position. It also aids the plastering process in that cables are securely fixed and supported before this process is commenced. Although metal capping is also available, as mentioned, it has been largely superseded by its PVC counterpart.

All-insulated sheathed wiring systems

The principal type of cable that the intruder alarm engineer will be familiar with in so far as the mains supply is concerned is PVC-insulated and sheathed cable. It is manufactured as single core (known as 'singles'), two core (known as 'twin') and two core with an uninsulated protective conductor (known as 'twin and earth') surface wiring cable. It is the last type that the alarm installer will be mainly involved with, together with three-core cable or cord that has three insulated cores within the outer sheath. These are used for surface wiring systems or are simply buried beneath the plaster or concrete, although in some instances mechanical protection may be needed. PVC cable will resist attack by most oils, solvents, acids and alkalis. In addition it is unaffected by the action of direct sunlight and is non-flammable, and is hence suitable for a wide range of internal and external applications.

PVC-sheathed cables can be run between floors and ceilings and dropped down through ceilings to spur outlets and such. Holes made for the passage of cables through ceilings can easily be filled with cement or another building material as a precaution against the spread of fire.

In so far as intruder alarms are concerned, there is still a need for the cable to be well protected, be it the mains cable for safety or the alarm cable for security. The entire system is to be protected from all likely damage, including mechanical, electrical and environmental. The interconnecting cables must be adequately supported and their installation conforming to good working practice.

Interconnecting wiring must not be run in the same conduit or trunking as mains cables unless they are physically separated. IEE Wiring Regulations do not permit the running of extra low-voltage cables with mains cables unless the insulation resistance of both are equal; however, it is usual practice to not run alarm cables alongside mains cables or even to run them through the same holes in building structures or to feed them through the same hole when entering an alarm control panel or power supply unit. There are special types of trunking available with separate compartments, and these can be employed for neatness and convenience.

In addition wiring is to be screened and protected from radio and electrical interference, which further enhances the need to keep the intruder alarm wiring separate from the mains cabling. Although the problems in the industrial and commercial sectors with regard to interference are much greater, this requirement applies to domestic practices also. Alarm wiring must also be kept separate from category 3 (fire alarm) circuits to ensure that they have the necessary resistance to extraneous environmental effects induced by category 1 (mains voltage) circuits.

The specific wiring requirements cited as a minimum standard is that alarm cables be multistrand four-core insulated and sheathed. These will be classified as hardwired and be essentially of 12 V DC type. Traditionally they may be referred to as communications cable, of very small diameter and using stranded conductor cores. This cable, used for interconnection wiring in intruder alarm systems, is as follows:

- The cable is only for use in internal locations and not for those areas exposed to the weather unless suitable protection is afforded.
- Cables must be segregated and shielded from interference.
- Conductors are to comprise solid annealed copper manufactured to BS 4109 Condition 0 with the strands to be plain or tinned surface coated.
- Solid conductors must have a minimum cross-sectional area of $0.2\,mm^2$ in line with BS 6350: 1988. If under $0.5\,mm^2$ cross-sectional area, the resistance is to be $95\,\Omega$ km for $0.2\,mm^2$ at 20°C and then pro rata to 0.5 mm. Stranded conductors should have a minimum cross-sectional area of $0.22\,mm^2$ to BS 6360: Table 3: 1981: Class 5.
- For cores comprised of conductors, solid or stranded conductors should be a minimum of 7×0.2 mm in diameter. Each core must be covered with type 2 or T11 PVC insulation manufactured to BS 6746 with a minimum thickness of 0.15 mm. The insulation must not adhere and it must also allow for easy stripping without causing damage to the conductors. The cores must be capable of withstanding a 1500 V RMS electric strength spark test. A cable is to comprise a number of cores of wires. Cables are to be sheathed with type 6 or T11 PVC manufactured to a BS 6746 with a minimum thickness of 0.4 mm and should include a rip cord to enable easy removal of the sheath without excess cutting that could lead to damage of the conductors or cores.
- The cable should be able to withstand an insulation resistance test of 500 V DC for 1 minute between each conductor and for all remaining conductors in the cable with an insulation resistance of no less than $50\,M\Omega$ for 100 minutes at 20 ± 5°C.
- Cables should be wound on to drums with an indication of the name of the manufacturer. The length of the cable should also be designated together with the number of cores, minimum cross-sectional area of the conductors in square millimetres, number and diameter of strands in a stranded conductor (e.g. seven at 0.2 mm), British Standard number and the date.

In practice, the intruder alarm engineer will be sourcing signal cable that will be generally specified as follows:

Maximum working voltage	60 V RMS
Maximum current	1 A per core
Maximum conductor resistance	92.4 Ω/km at 20°C
Maximum operating temperature	70°C
Conductors	7/0.2 mm strands of annealed copper wire conforming to BS 6360
Insulation	PVC radial thickness 0.2 mm nominal, confirming to BS 6746
Sheath	PVC nominal wall thickness 0.5 mm
Nominal overall diameter	4 core: 3.5 mm
	6 core: 4.1 mm
	8 core: 4.5 mm
Wire insulation colours	4 core: red, blue, yellow, black
	6 core: red, blue, yellow, black, white, green
	8 core: red, blue, yellow, black, white, green, orange, brown

This cable will be advertised generally as general-purpose four-, six- or eight-core signal cable ideal for use with security alarm systems and other similar applications where low voltages and currents are employed. The cable contains flexible wires each having seven strands of 0.2 mm tinned annealed copper insulated wire.

This cable will of course satisfy the vast majority of wiring functions within the intruder alarm sector covering the detection devices and signalling equipment, although connections to remote signalling components employing British Telecom (BT) approved equipment must use approved and type-tested cable.

It is possible to obtain cable to the same specification but with an even greater number of cores, of which 12-core cable is an example, but because of the greater use of multiplex bus wiring and addressable alarm systems we can see a move to conductors with fewer cores to achieve the same end-result. There will also be moves in the future towards high-grade shielded transmission cables.

The BT type-approved cable for connection of remote signalling equipment must conform to the requirements of BT specification CW1308. This cable has 1/0.5 mm tinned annealed copper conductors (single core) and is covered with 0.15 mm PVC insulation and an overall outer PVC sheath. It has a voltage rating of 80 V AC at up to 10 kHz and an insulation resistance of 50 MΩ at 20°C. The identification system is different for BT type-approved cable, as it uses a pair method of coding in that the base colour of a core of a pair is the band colour of the other core, thus white/blue and blue/white through to orange and green for a three pair (six-core cable).

Cable formats and considerations

At this point we can go on to look at the general constructional format of the principal cables in which the intruder alarm engineer is interested. We can then overview the role of twisted pair cables used in data transmission and the processes we use for general selection.

Conductor type

The most popular material for the actual conductor medium is copper. In relative terms, it has low electrical resistance, is not unduly affected by corrosion, is not heavy, has good qualities of strength and is economic. Gold, silver, steel and aluminium cannot compete in the overall scheme of things.

The conductor is available as either stranded or solid core. Stranded cable is the more flexible and can be easily installed and routed through awkward points. It is therefore used to great effect as the predominant cable in all mainstream intruder alarm systems and for patch cables around equipment.

Stranded cables do suffer from greater power losses than solid core as the current flows through the conductors. However, this is not a consideration in intruder signalling or in those systems which only involve low values of voltage or current.

Solid core is a better conductor for long runs and if attenuation (power loss) is seen as a distinct problem, but it can harden and break if heavily flexed or continuously adjusted. It is used in many telecommunication applications and for general fixed wiring.

It should be noted that as the temperature of the cable increases, the resistance of the conductor also rises. For this reason cables must be protected when running through areas that can be subject to high ambient temperatures.

Cable construction

The cable in its simplest form is a copper conductor held within its length by an insulator. But security cables carry data and when connected to sensitive equipment need additional features added to counter the effects of external interference.

Because the changing currents in other cables in the same general area create a changing magnetic field, this can be induced as electromagnetic waves in the form of electromagnetic interference (EMI) into our security cables. EMI in our system distorts the current flow and can result in system crash or memory erasure.

For this reason, specified cable has the actual conductors insulated and wrapped in a conductive medium of copper or aluminium foil, metal tape, braid or even a sleeve. This protective medium shields and protects our cable from EMI.

At the time of installation, by connecting the shield to a good earth point, the EMI that may have been induced into it is able to leak away. Often a copper drain wire is run with foil along its entire length, as it provides an easy way of making the earth connection at the cable ends.

A plastic or fibre rip-cord may also be included to help the jacket to be stripped back. Although EMI is the main problem encountered, the foil or braid is also able to give a measure of protection against other induced electrical fields.

Twisted cable

For standard signalling cable, the conductors run parallel with each other throughout the cable length. But many specifications now call up twisted cable.

These are available in an enormous range of variants. Twisted pair cables are twisted along their length, with one of the pair conductors carrying the signal and the other operating as the return path.

The twisting of wires greatly reduces the effects of EMI, as the interference induced into it acts in opposite directions. Copper unshielded twisted pair (UTP) is presently the dominant cabling medium in building data networks.

Data transmission

In regard to the sending of data, telecommunication speeds have increased enormously over the last few years and will continue to do so. The bandwidth of a system governs how fast information can be transmitted over the communication links, and the cable has a major role to play in coping with the speed of this data transmission.

The cables we use are known by the Categories 1 to 7 or the international groups, Classes A to F, with 1 and A being at the lower end of the scale in terms of bandwidth.

In practice, the most popular cables used are Category 5e (enhanced) and Class D, as the lower categories are obsolete or relatively 'slow'.

Taking this a stage further, Category 6 copper cabling systems are claimed to be compatible with fibre optic links in terms of speed.

Installers are involved with a huge variety of cables, including electrical installation, flexible mains, data transmission, audio/visual, signal, telephone, co-axial and screened. Although certain systems are clearly defined in terms of acceptable cables such as fire detection and emergency lighting, not all of the systems of interest to the intruder alarm installer are so established. Therefore cable selection must take many aspects into account. During the fitting of cables the alarm installer must ensure that there is segregation of the security cables from the other wiring services.

Table 7.1 gives an idea of what needs to be taken into account in the selection process.

Table 7.1   *Cable selection process*

Considerations	
History	History and performance of the proposed cable type in a similar installation. Compatibility with other cables in service.
Inspectorate requirements	Acceptance by regulatory authorities. Industry codes of practice may govern and control acceptable cables.
Future developments	Upgrading and expansion of systems at a future date. Bandwidth may need attention. Future integration with other building management systems and networks.
Power requirements	Cater for power loss in long runs and attenuation (reduction in signal strength at a specific frequency). Estimates to be made for length of cable runs.
Bandwidth	Selected cable needs an acceptable bandwidth to satisfy the rate of data transmission.
External fields	Need for shielding and twisted pair. Possibility of electrical interference being induced by other building services. Processes generating discharges and inductive loads need attention.
Mechanical protection	Cable may need a special sheath to suit the environment. Containments such as conduits/trunking may need to be applied. Cable may need additional measures to combat tampering.

7.2 Installation of supports and cables

In view that the intruder alarm engineer is mainly interest in PVC-sheathed wiring we will only focus our attentions on the supporting of this cable form being the mainstream practice. PVC-insulated and sheathed cables must of necessity be protected from mechanical damage in certain areas, and this has been considered. However, in some areas this is not essential as the cables may not be accessible or be concealed. They do, however, still need to be supported. The cables must be fixed at intervals as listed in Table 7.1.

Table 7.2 refers to cable supports in accessible positions, and this is by means of clips or saddles. PVC clips with a single hole fixing are most popular, with the internal surface of the clip being formed to suit the cable size and form. Self-adhesive clips can also be used where it is not possible to drive the nail fixing into position.

In the event that cables are installed in normally inaccessible positions and are resting on a reasonably smooth horizontal surface, then no fixing is necessary. However, fixing must be provided on vertical runs over 5 m.

When running in cables it is important to apply incombustible material to holes to prevent the spread of fire. Tubing should be used when negotiating sharp surfaces, including brickwork. Cables under floors must not be installed so that they can be damaged by contact with the ceiling or floor or their fixings. Cables passed through drilled holes in joists must be at least 50 mm vertically from the top or bottom and be supported by battens over extended runs.

Where cables pass through structural steelwork, holes must be fitted with suitable bushes to prevent abrasion.

The only other considerations are apparent, and require that care be exercised when removing the sheath of the cable and that this should be of the correct length to allow cables to be pushed back into position after terminating without too much tension or too much slack. Conductor ends must not be reduced as this leads to a loss of current-carrying capacity, and clamping screws must be well tightened.

Table 7.2 *Spacing of cable supports for PVC-insulated cables in accessible positions*

Overall cable diameter (mm)	Horizontal (mm)	Vertical (mm)
<9	250	400
>9 but <15	300	400
>15 but <20	350	450
>20 but <40	400	550

We can conclude our observations on security wiring requirements by looking at their integration in Europe and summarizing some details. Cables are to follow contours. They must not be closer than 10.5 cm to any fixing point, such as the corner of a ceiling to wall, and be sited away from door frame uprights. They must also travel in straight lines and never diagonally across walls. They must be protected if likely to suffer damage, and when passed through roof spaces and between joists they should be encased in high-impact PVC or metal conduit. This is to reduce line faults and tampering potential. Jacketless cables are not to be used, and they should be sleeved in trunking with the same resistance to fire as the building material when passing through a floor or wall material or when buried. Cables are not to pass near steam or hot water pipes, and are to be at a reasonable distance to prevent any rise in surface temperature above the designed ambient temperature of the jacket.

Discussion points

Having considered Sections 7.1 and 7.2 and the options when considering an appropriate wiring system, the installer must take a decision on the best practice with all factors taken into account. This ranges from protection to aesthetics for both mains and signal cabling. The decision to use steel conduit with its electrical continuity or an alternative non-metallic type of protection is important and will certainly differ between environments.

Once this has been established the consideration is that of connecting and terminating cables, and this forms the next stage.

7.3   Joints and terminations

Having studied the wiring requirements for security systems, the need to join and terminate cables necessitates consideration. We know that installing multiple sensors of any type on a single detection circuit will increase the load and lack of cover in the event of a fault. We also realize that it makes fault finding more difficult. It is for this purpose that circuits are provided with an adequate number of test points in accessible positions, although they are best concealed. This clearly requires joints or splices, with an acceptable device being employed or the connections being soldered. However, there must be no identifiable markings on the cable sheath and no labelling should be applied although random colour coding should be used with the details held securely by the installation company. Acceptable methods for joining wire are:

- wrapped terminal joints and splices;
- crimped joints;

- soldered joints;
- clamped joints.

Great care must be used with joints for two specific reasons:

(1) system resistance is increased and voltage drops can be caused with an inherent reduction in reliability;
(2) points are introduced where the system can be subject to abuse by bridging and tampering.

The normal interface between the electronic designer and the installer can be viewed as the wiring connections, be it a plug and socket, barrier strip, terminal block, tab connector, wire wrap, crimped sleeve or solder joint. Indeed, it is at this interface that so many installation and service problems originate. Copper is of course the normal material for interconnecting wire, but it can easily be damaged if excess screw pressure is placed upon it. Signal cabling is made up of several strands of fine-gauge wire rather than a single strand of heavier gauge. BT cable and mains supply cable, however, normally comprise a single conductor, and the technique of terminating it will be slightly different. Although we can provide the student with a description of the methods and practices that are available, in practice his or her ability to produce good connections will very much be governed by experience.

Mains connections

The cable to the intruder alarm is always best as an unjoined run derived from the consumer unit. However, there are junction boxes that can be used if the need arises. These must be placed in accessible positions and suit the load and cable. The sizes, current ratings and principal circuits are given in Table 7.3 for the domestic environment. The junction boxes will be of circular form and of 5, 15, 20 or 30 A capacity with three or four terminals of brass construction. They are often found in systems using twin and earth cabling.

Table 7.3 *Sizes, current ratings and principal circuits. Mains connections*

Cable size (mm²)	Current ratings (A)	Circuits
1.0	16	Lighting
1.5	20	Lighting and 15 A single sockets
2.5	28	Ring circuits and 20 A radial circuits
4.0	36	30 A radial circuits

The method of connecting is simple in that like conductors are clamped together under the same terminal screw with earth insulation being applied over the protective conductor. The insulation for the phase and neutral conductors must only be stripped back as far as necessary, and the outer sheath must enter the junction box slightly before being stripped back. The important thing to remember about the supply connection as opposed to signal cabling is that with the former any bad or loose joint will lead to sparking, high resistance and the generation of heat that will eventually result in the destruction of the insulation. This can ultimately lead to fire or electric shock.

Intruder wiring

The total resistance of all the circuits must be less than that which would reduce the voltage to below the required minimum at full load. The standard permits joints that are wrapped, crimped, soldered, clamped, connected by plug and socket or wire-to-wire and are either insulated or in a junction box. All wiring must be within the protected area, or where this is not possible then it must be mechanically protected.

The regulatory authorities require wiring to be protected against damage or tampering and for it to be of a neat and professional appearance.

Circuit wiring should be identified so as to facilitate future fault location with colour coding, but this can be varied with the code noted in a record book, and voltage drops on long runs must be considered. Junction boxes are to be fitted in complex circuits to allow circuit faults to be easily traced and be of an anti-tamper variety in high-risk situations. Open wiring can be secured by means of cable ties or by insulated staples, but runs are to be along the sides and top of architraves and in the lip of a skirting board or picture rail to avoid any damage being inflicted on the woodwork. Staple guns can be used but treated with caution to avoid damaging the cable insulation.

The engineer is therefore given some freedom with respect to the method of jointing to be employed, and can decide which method is best used in any given situation. Soldering requires technique, and is discussed separately.

Soldering

Soldering irons range from small gas-powered devices through to miniature soldering irons either powered by 12 V batteries or from the mains 240 V supply with ratings from 15 W up to the higher-capacity 50 W soldering stations.

The most commonly used solder is referred to as 60/40, being an alloy of 60 per cent lead and 40 per cent tin. Its melting point is low enough to allow safe soldering of most heat-sensitive electronic components, and is used with a non-corrosive flux that automatically cleans away oxides formed during the soldering process. It is available in a solder reel pack in different gauges to suit the parts to be soldered. The melting point is of the order of 190°C, and the gauges will be generally 18 SWG (1.22 mm) and 22 SWG (0.71 mm).

To actually perform the soldering process there are a few time honoured tips:

- 'Tin' the soldering iron tip by melting a little solder on to it after first cleaning the surface area.
- Ensure that a mechanically strong joint has been obtained by twisting wires together so that when the solder is applied it will lock and seal the surfaces.
- Hold the iron to the joint for a short period to preheat the joint and add the solder to the joint. If the solder does not melt, withdraw the soldering iron and apply more heat to the wires.

A good solder joint is shiny and smooth. A bad 'cold' joint is dull and may also be rough. When joined wires are soldered, the solder should flow with a smooth contour to meet the wire at the ends of the joint; however, if it forms a blob with thick rounded ends the joint is not satisfactory.

An excellent level of mechanical strength can be obtained by a good soldered joint together with good electrical conductivity. The installation can then be complemented by the application of sleeving being drawn into position. To this end, heat-shrinkable sleeving can be applied which will shrink to a given percentage of some 20–50 per cent of its original diameter after the application of heat by a hot air blower or such. These sleeves have a high dielectric strength, resist attack by solvents and alkalis and can even provide a moisture-proof seal.

Crimping

The question of whether this is a better method of joining than soldering remains unanswered but we do know that a cold solder joint will not provide adequate electrical and mechanical characteristics. The engineer will need to use his or her own judgement. Certainly stranded wire crushes down well, so crimp-on connectors are held firmly. Crimping tools used correctly with the proper pressure being applied will provide good electrical and mechanical connections. There is a good range of crimping connectors available, and the task of applying them is certainly quick and easy.

Soldered and crimped connections are made up within tamper-protected junction boxes, but there is also a range of proprietary junction boxes specifically available for connecting cables. These boxes are in PVC with an anti-tamper lid, have countersunk back holes for fixing and screw terminals with which to clamp the wires being joined. They are available as 6, 8, 12, 20 and 24 way variants, and have break-outs for the cable entries. An addition to the range is the door loop version with its flexible integral silicone-insulated cord to allow placement of the wiring into moving doors.

Clamped joints

The considerations applying to the use of screw wrap terminals are apparent. Use special wire strippers to remove the cable insulation. After stripping, twist the bared wire in the same direction as its natural twist to prevent any stray strands. Use only a clockwise wrap under the screw heads making sure that only some 1 mm of bare wire is exposed beyond the screw head. Always support the wire when tightening any screws on to them. Ensure that the conductor is firmly gripped but not overtightened or damaged. Do not let the outer sheath be removed outside of the junction box, and ensure that the correct amount of slack is allowed within the enclosure and that the wires are not trapped by the lid or are pulled tight when the lid is resited. These considerations are very elementary and do not need elaboration, but when correctly performed, good electrical characteristics and mechanical strength are achieved.

There are of course a huge variety of other forms of clamp, but the normal procedures apply with all screw forms whether the screw is applied direct to the joint or applies a spring leaf against the connection.

Splicing

The definition of the splice, is that of a joint essentially made by tightly winding the ends of two wires together for a short distance and then mechanically restraining it by an acceptable method such as soldering, crimping or sleeving.

Discussion points

With the progression towards more complex wiring and in cases of longer wiring runs there is a need to apply joints to facilitate fault finding. The provision of any joint creates a multitude of considerations, and the engineer must balance these against his or her capacity to perform satisfactory connection techniques.

7.4   Fixing methods for devices

The method of fixing is very much dictated by the device that is to be installed and the material that it is to be affixed to. In many cases manufacturers supply fixing devices such as plugs and screws to guide the installer, and this is very much the case with most external sounders. These of course have their own requirements and can be considered at the outset.

External audible intruder alarm signalling devices

The manufacturer will often provide guidance, but these devices should be fitted as high as possible to reduce the possibility of interference by an intruder. The position may also depend on a number of factors which can affect the ability of it to be heard (or seen in the case of a visual warning device). It therefore should be located in a position where it may be easily viewed. The device should not face heavy traffic or a railway as this will affect the ability of it to be heard at a relatively short distance. The wiring to it should enter direct through the wall and not be surface mounted.

System alarms are divided into bells and sirens, although bells are certainly going out of style. The construction of external audible alarms is governed by the standards applicable to the extent that they must be totally enclosed and weatherproof with a case offering at least the same protection afforded by 1.2 mm of steel. They must also produce at least a 70 dB(A) mean sound level and 65 dB(A) in any one direction at 3 m with the security cover in place. These sounders are liable to be pulled from walls, and therefore must rest on a solid structural base of concrete, brick or hardwood, which will also prevent some of their power from being absorbed through vibration of the building medium.

Therefore, they must be sited to give maximum prominence and sound output yet have reasonable protection from accidental or wilful damage but still provide access for servicing.

The surfaces to which an external sounder may be affixed will vary enormously, and the minimum requirements for typical surfaces are described below.

Brick walls

Three No. 10 screws in suitable plugs penetrating the actual brick to a depth of at least 40 mm should be used.

Metal, wood or thin skinned structures

Bolts with back-plate are required for these surfaces.

The first type of fixing is straightforward and there is a huge range of screws and plugs available. The screws are available with different head forms and finishes and even with high-tensile characteristics. In installations where the wall structures are irregular it is advisable to seal around the enclosure back-plate with cement to make it more difficult for the intruder, who may attempt to insert a jemmy behind the sounder in order to prise it from its fixings.

The second type of application is encountered on prefabricated buildings. In these instances it is normal practice to use a hardwood back-plate which is initially fixed to the building structure using bolts with washers and nuts. The cable is brought through this back-plate via grommets, and the sounder is also bolted through the back-plate and building medium. When carrying out this task the back-plate is best fixed in place initially and then further bolts used to secure the sounder rear housing. The sizes of the bolts and back-plate depend on the application and its vulnerability, but, as a minimum, a sheet of 20 mm hardwood with 6 mm bolts, steel washers and steel full nuts or nylon insert locking nuts is practical. The heads of the nuts will be protected by the sounder enclosure to stop any attempt at removal.

Remote signalling equipment

Remote signalling equipment is required to be within a secure area that is not visible from outside of the building. Its housing must have the same resistance to attack as a control panel and have appropriate anti-tamper devices to prevent it being removed from its fixing without generating an alarm. The housing must be securely fixed in position. Due to these requirements these devices are often installed within cupboards, but such areas may not have solid walls and can often be of cavity construction.

So what are the different wall types that must be accommodated?

Solid walls

General-purpose wall plugs can be used with appropriate screws (Table 7.4). Where more heavy duty fittings are required, heavy duty wall plugs intended for brick, concrete, breezeblock and cellular block or stone

Table 7.4   *Plug and screw fixing sizes*

Hole diameter (mm)	Screw size
5	Nos. 4–7
6	Nos. 6–10
8	Nos. 9–14

should be considered. These will use, in general, a 7 mm drill bit and accommodate screw sizes Nos. 8–14. Dry-lined walls also fall into this category (plasterboard attached by spot adhesive to a solid backing wall).

Partition/cavity walls

The position of the wooden framework should be established, and long screws used to fix equipment to this structure. Heavy duty plasterboard plug fixings can be used as a supplement. These are specific plugs intended to fix to cavity and plasterboard partitions. Alternatively, super toggle cavity anchors can be adopted. These have screw-tensioned nylon anchor arms which self-adjust to the wall thickness and provide a strong method of fixing for cavity walls. For an even more superior mounting, cavity fixing spring toggles can be used. These are designed for use with plasterboard, lath and plaster and other partition materials. These feature a strong grip and have wide span wings, and are installed as shown in Figure 7.1. Bolt sizes can be up to M6 (6 mm).

Stone walls

A strong fix to stone walls can be made using general-purpose wall plugs or the heavy duty variant. The main consideration is to ensure that the plug is entering stonework and not loose or soft surrounding plaster or cement. Often, longer screws are needed to reach the stonework because of irregularities in the wall.

The conclusion to fixing is to remember that the wall medium must be suitable and that the screws of the correct length are used to reach far enough into any plugs to ensure that they cannot be pulled from their location.

Figure 7.1 *Cavity fixing spring toggle*

7.5   Working equipment: safe use

The Health and Safety at Work Act 1974 (HASAWA)

The very first health and safety legislation in the UK was passed as early as 1802. Over the years, a great many pieces of legislation were passed, although much of this was very confusing. This was clarified in 1974 when the HASAWA was introduced. Its first major significance was that it covered everyone at work whatever the workplace with the exception of staff in domestic premises. The Act placed clear duties on everyone, employer, employee, the self-employed, manufacturers, suppliers and installers.

The main purpose of the Act is to promote good standards of health and safety, so preventing people coming to harm at work. It makes health and safety an essential part of work, not an option. It does this by placing statutory duties on employers and employees to build into their work safe practices. It does not give step by step instructions on how to do this, but details can be found in regulations made under the Act. Examples of this are the Electricity at Work Regulations 1989 and the First Aid Regulations 1981. The Act is proactive and not reactive, and the framework that it sets up allows for the ongoing process of developing health and safety legislation by updating older regulations and issuing new regulations, some of which originate from the EC.

Most of the responsibility for health and safety falls on employers. They must ensure the health safety and welfare of their employees by:

- providing safe systems of work, safe environments and premises with adequate facilities;
- provide safe access and egress to and from the workplace;
- provide appropriate training and supervision;
- provide information to the employees;
- have a written health and safety policy if there are five or more employees;
- provide safe plant, machinery, equipment and appliances and safe methods of handling, storing and transporting materials.

Employers must also make sure that their activities do not endanger people who visit the workplace or members of the public.

Employees have a duty to take care of themselves as well as anyone else who may be affected by what they do at work. They must also cooperate with the employer on health and safety matters by following rules and procedures.

The self-employed are covered in much the same way, with a duty to ensure that they do not endanger themselves or others by their work activities. The installer working in a private house is governed by the same rules that would apply in a large industrial building, and he or she must therefore:

- design and construct a safe product;
- test the product for safety;

- provide information and instructions for the user;
- ensure that the product is safely installed.

In so far as safety is concerned, a lot is down to common sense, and accidents can be avoided by being aware of hazards and by following established rules and practices. Some human factors that may cause accidents are:

- carelessness;
- inexperience;
- lack of training;
- haste;
- distraction;
- complacency;
- influence of drugs or alcohol;
- breaking safety rules.

It follows that an employer has to ensure that persons are properly supervised at work, are trained correctly and understand safety procedures. Some tasks have obvious hazards, and within the intruder industry working at a height on poorly maintained access equipment or working with poorly maintained power tools will always create problems.

Ladders

Figure 7.2 shows a set of ladders tied securely and at the correct angle. It has been found that more than half of the accidents involving ladders occur because they are not securely fixed or placed, and indeed most of these accidents occur when the work is of 30 minutes duration or less. It is apparent that most of this is down to haste. Certain considerations apply when using ladders:

- Support the foot on a firm level surface. Never on loose material or other material to gain extra height.
- Do not tie shorter ladders together to obtain the desired height.
- When possible, secure the top of the ladder using clips, lashings or straps.
- Alternatively, secure at the base using blocks, cleats, sandbags or stakes embedded in the ground. Note that 'footing' is not considered effective for ladders longer than 5 m.

The safe use of ladders also involves the following:

- Make regular inspections of them.
- Never carry out makeshift repairs.
- Never paint wooden ladders as this can hide defects. Clear varnish may be used.

1.05 m (3ft 6in) or over

Tie securely

4

75°

1

Figure 7.2   *Ladder position*

- Beware of making contact between aluminium ladders and live electrical cables.
- Ladders should extend at least 1.05 m above the highest rung being stood upon.
- The angle of use should be 75° to the horizontal and 1 m for every 4 m in height.
- If the ladder is erected near a doorway it must face the door, which is to be locked shut or secured open. The base must also be fenced off to prevent others walking under it.
- Never allow more than one person at a time to be on a ladder.
- Never consider the ladder as the best device for all applications: it may be safer to work from a securely constructed scaffold tower.

Tools and lifting materials

Do not carry heavy items up a ladder. Use a rope or hoist. Carry tools in a bag or belt holster to enable the free use of both hands to secure a firm hold.

Stepladders

These and trestles must never be used for any degree of side loading. Many injuries and fatalities have occurred when descending from work platforms and landings using unsecured stepladders. Avoid overreaching.

Storage

Store wooden ladders away from radiators, steam pipes and other sources of heat. Store all ladders in a dry and cool place and support them only on the lower stiles by rack or on wall brackets. Never allow them to be hung by the rungs.

8 Inspection and testing of the low-voltage (mains) supply

The Electricity at Work Regulations requires that people who conduct these tests are competent under the terms of the Health and Safety at Work Act. This chapter provides students with instruction on how these tests are safely carried out and how to achieve competence.

This chapter discusses the need to visually inspect the installation and studies the methods of testing the low-voltage supply to the system, covering earth continuity, polarity, insulation and earth loop impedance. There are certain safety precautions to be observed, and procedures for recording the test results using proprietary or commercial forms need to be followed.

8.1 Part P of the Building Regulations

Historically the Electricity at Work Regulations have required that people who work with the low-voltage (mains) supply and carry out the inspection and testing are competent under the terms of the Health and Safety at Work Act. Therefore all of those involved with the installation and maintenance of intruder alarm systems have always been committed to be able to demonstrate competence. However the inclusion of electrical wiring in the Building Regulations in the form of Part P came as a Government initiative launching new rules for electrical safety in the home and as a means of enshrining it within the law. The Government's Office of the Deputy Prime Minister (ODPM) welcomed the implementation of Competent Persons assessment schemes in the industry to be run by a number of recognized providers. It was through these schemes that contractors must apply for an approval status.

The Building Regulations which are approved by Parliament contain a schedule of requirements that are referred to as Schedule 1. This Schedule contains a number of Parts each of which is supported by related Approved Documents. These Documents reproduce the requirements and give practical guidance as to how they may actually be complied with.

Part P of the Building Regulations for England and Wales came into force on 1 January 2005 to cover electrical safety. The implications of these new statutory requirements was wide-ranging and had a distinct effect on all those undertaking work on electrical installations in a dwelling including DIY persons. This was because from that time they must either be registered under the recognized assessment scheme as competent to certify their work or alternatively they must inform the Local Authority Building Control who must then subsequently inspect and certify the completed works. All electrical installation work in dwellings, that is generally houses and flats, became a controlled service under the Building Regulations. Therefore for the first time the technical standard of electrical installation work in dwellings became subject to statutory requirements. Such requirements applied not only to new constructions but also to alterations or additions to existing installations, including full or partial rewiring. In Scotland the introduction of the Building (Scotland) Regulations 2004 Part N became standard 4.5 with changes in terms of the need for warrants but with the technical requirements being unchanged.

In basic terms Part P was introduced to ensure that all UK electrical installations in dwellings complied with the IEE Wiring Regulations by introducing the need for new work to be certified by a competent person. Part P in practice covers both small scale domestic work for consumers and large scale developments for businesses. Therefore it had an impact on contractors at all levels.

The local authority building control was given the task of enforcing the regulations and ensuring that all UK electrical installations in dwellings complied with the 16 edition of BS 7671, The IEE's Wiring Regulations. Demonstrating compliance involves carrying out tests as prescribed in BS 7671 and recording the results. However, there are certain installation practices that do not need to be notified to the local authority, i.e. the person involved does not need to be approved as a Competent Person (CP) to carry out work within the following groups:

- All replacement work within the dwelling to include the replacement of ONE cable damaged by impact, fire or a rodent. This does not include the replacement of a consumer unit.
- Providing mechanical protection to cables. This covers trunking and conduit etc.
- Installing additional socket outlets, fused spurs and lighting points etc. that are outside of the 'special locations'. Such 'special locations' include kitchens, bathrooms, conservatories, and outside of the property.
- Installing or upgrading of earth bonding.
- Any circuitry or equipment that runs on 50 volts or less, i.e. telephony, aerial and satellite wiring or intruder or fire alarm cabling. This all falls outside of the scope of the IEE Wiring Regulations BS 7671: 2001 so is exempt from Part P.

It follows that, although the intruder alarm installer may not be required to become approved as a competent person, the work performed with the low-voltage mains supply is limited to the aforementioned groups.

However, because of the greater integration of intruder alarm systems with other building services and safety and security networks, the intruder alarm engineer has become increasingly involved with mains supply volt-ages and additional power supplies and mains driven components.

Nevertheless, even if there is no specific need to become approved as a competent person the contractor must still be able to demonstrate compe-tence, and as the installation is within the requirements of the IEE Regulations there remains a need to work to the demands of that standard. This includes the compilation of a 'Minor Electrical Installation Works Certificate' for the end user. Equally everyone has a duty of care under The Health & Safety at Work Acts 1974, The Electricity at Work Regulations 1989 through to The Provision & Use of Work Equipment 1992.

Therefore it follows that the intruder alarm engineer is committed to understand and be able to work within the scope of the Wiring Regulations for the system parts which are involved.

It should also be noted that it is installations that have to comply with Part P and not the instruments used to test it, and that only three basic instruments are needed for testing domestic electrical installations, testers for: earth loop impedance; insulation and continuity; and residual current devices. As an option proprietary instruments are now available as Part P kits with menu driven design facilities. In all cases properly calibrated instruments are needed to satisfy BS 7671.

8.2　Visual inspection and testing

Most intruder alarm systems feature connection to the low-voltage mains supply, and the installer must therefore fully understand the requirements that apply and the safety precautions that exist.

After completion of any new wiring installation or alteration to an exist-ing system, the work must be inspected and tested to ascertain that there are no defects and that all necessary conditions have been satisfied. However carefully an installation has been completed it is always possi-ble for faults to occur at a later stage by nails being driven into cables, insulation being damaged and connections being broken or defective apparatus being installed.

A full test of the completion of the mains connection by the intruder alarm installer should be carried out and results carefully recorded. The local electricity board is entitled to refuse to connect a supply to a consumer if it is not satisfied that the installation complies with the relevant statutory regulations. In the UK an installation which complies with the IEE Wiring Regulations is deemed to meet the statutory requirements and should there-fore qualify. The procedures for inspection and testing on completion of an installation are covered in the IEE Wiring Regulations, Part 7, which includes a checklist for visual inspection. Following the inspection, tests should be performed on the installation.

8.3   Safety precautions and test equipment

Regulations make it an offence in so far as electrical connections are concerned for a person to make permanent connection to any part of an electrical installation unless competent to do so. This is in accordance with BS 7671 and the IEE Wiring Regulations. If anyone other than a competent person installs the mains connection to any system and an accident or fire occurs as a result, then both the customer and the person that carried out the work are legally responsible. These wiring regulations have been in existence for over 100 years. They became a British Standard in 1992 and were published as BS 7671.

The intruder alarm and power supplies must be mains connected so that they comply with the Electricity at Work Regulations and the related BS 7671 standard.

The installation must also be such that connection is to the mains supply via an unswitched spur unit. This must use at least 0.75 mm cable with the spur sited adjacent to the equipment. It must essentially be within arm's reach of the equipment so that it can be isolated before the control panel is opened.

The fuse should be appropriate, and 3 A is typical.

Intruder systems do not specifically call for the spur to be direct from the consumer unit, but it is important to ensure that it is not possible for someone to inadvertently isolate or restore power from another source. It follows, therefore, that it is best to run the spur for the system direct to the consumer unit. If this is not possible, any other master spur should be marked to indicate that it also serves the security system. Drawings and diagrams should be updated to that effect.

The intruder alarm engineer should never work on any electrical apparatus that is live to the mains supply, and must power down the system before any modifications are made. Of necessity this applies when adding or changing detectors or printed circuit boards, when it is also necessary to disconnect the standby batteries and silence the sounders – beware if a non-volatile memory (NVM) is not fitted because programming will be lost so a manual would be needed!

Perhaps before we consider safety precautions we should look back a stage to how the AC mains supply is actually derived.

AC mains supply

Power supplies are generated almost everywhere as alternating current (AC), which means that the current is changing direction continually. In the UK this change of direction occurs 50 times per second (50 Hz). The machines that actually generate this power are known as generators, and they have three identical sets of windings in which the current is generated.

One end of each winding is connected to a common, or star, point which is termed the neutral. The other ends of the windings are brought out to the three wires or phases of the supply cables. For identification purposes, these are colour coded as red, yellow and blue, and the currents which are transmitted in each phase have a displacement of 120°.

Power supplies to towns and villages are provided from power stations all over the country, via a system of overhead and underground mains and transformers which reduce the voltage in steps from the high transmission voltage to the normal mains voltage used by the consumer. The National Grid is responsible for the transmission of power in bulk, at 400 kV, on the Super Grid System. The local electricity boards are then responsible for the distribution of power from the grid stations to all the industrial and domestic users. This is achieved by using a network of overhead lines and underground cables (usually at 132 and 33 kV) to take the power to the main load centres, and 11 kV overhead lines and underground cables to distribute the power to individual load centres, where secondary substations reduce the voltage to mains potential. From these substations, which consist of pole-mounted transformers and isolating gear or ground-mounted transformers and switch gear, low-voltage overhead and underground mains are taken to the consumer's supply terminals. AC power is distributed mainly on three-phase networks, although in certain rural areas only single-phase supplies are available. Thus, on the low-voltage side of most local transformers will be found four terminals: the red, yellow and blue phases and the neutral. Between each phase and the neutral there is a voltage of 240 V. However, between the phases the voltage is 415 V. For this reason the secondary output voltage of transformers was traditionally given as 415/240 V. These values are still often quoted although we now see 230 V applied as the single-phase mains voltage. This is illustrated at Figure 8.1.

The supply authorities are required to maintain the voltage at the consumer's supply terminals within 6 per cent of the nominal voltage, which equates to 224–256 V. The various systems for distributing power to consumers are defined in Part 2 of the IEE Wiring Regulations, and in the UK are designated TT, TN-S or TN-C-S. Most domestic customers are provided with a single-phase supply, unless exceptionally heavy loading is expected. Most industrial premises, however, are provided with a three-phase supply since the load may be high and heavy machinery requires three-phase motors.

Power transformers are more efficient if the load on each phase is approximately the same, so single-phase services are normally connected to alternate phases, and three-phase consumers are encouraged to balance their loads over the three phases.

The local boards provide underground cable or overhead line services which terminate at a convenient point within the premises. Overhead services are terminated on a bracket high up on a wall of the property, and

Figure 8.1   *Three-phase supply*

insulated leads taken through the wall to the meter position. Underground services tend to be brought through below floor level via ducting. The overhead or underground service leads are taken into the main fuse of the electricity board. This is usually 100 A for domestic purposes. From this fuse point the supply is taken to the meter; however, when three-phase supplies are provided, three main fuses are used with one composite meter.

With underground services, either the lead sheath of the cable or, as in the case of plastic cables, the wire armour is used to provide an earth. A separate wire is generally bound and soldered to the sheath or armour at the terminal position and then taken to an earth connector block. All protective conductors in the property are taken back to this block.

With overhead services, an earth block may be provided if protective multiple earthing (PME) is adopted. In other cases an earth electrode, in conjunction with a residual current circuit breaker, must be used. In some cases a separate overhead earth conductor is provided. In large blocks of flats or offices the services to each floor are provided by 'rising mains'. These will not normally be encountered in domestic work. In such cases where the intruder alarm engineer finds rising mains, it will be seen that consumer's meters are located on the individual floors, and the rising mains are used to carry the bulk supply up the building. Subservices will branch off at the various floor levels. The mains themselves consist of one of the following systems:

(1) Rigid conductors in a protective enclosure.
(2) Single-core or multicore, paper-insulated or PVC-insulated and sheathed cables or mineral-insulated copper-sheathed (MICS) cables. These are run on cleats or a cable tray in a vertical chase which must be free from combustible material. Multicore paper or PVC cables are normally armoured.
(3) PVC-insulated, single-core, non-sheathed cables enclosed in conduit or trunking.

Consumer's meters are often arranged so that they can be read without a need to enter the premises. The meter may be visible through a small vandal-proof window. In multistorey blocks of flats the meters may be fitted in the riser duct running up the building or other communal area. It follows that the intruder alarm engineer should be aware of where the supply originates and enters the building. It also follows that there may exist more than one phase in a given premises, so there are extra precautions that need be taken to ensure that work is not being performed across more than one supply. These multi-phase systems tend not to be encountered in the domestic environment but *they certainly will be* in the commercial and industrial sector. At all times ensure that the supply has been inspected and tested as previously described. It now becomes appropriate to look at the protection of circuits and apparatus.

Protection of circuits and apparatus

It is essential to ensure that all fixed installations or portable apparatus are correctly protected against overloading and also that there is no risk of fire in the event of a fault developing.

Fuses

A common method of protecting circuits and apparatus is by the use of a fuse in line with the phase conductor. The fuse is essentially a device for protecting a circuit against damage due to excessive current flowing in it. It operates by opening the circuit when the fuse element melts because of the heat it has been subject to, when current above its rating flows through it. The two main types of fuse are:

- *Semi-enclosed (or rewireable)*. These have a fuse holder or link of an incombustible material such as porcelain or moulded resin, and a fine wire between the two contacts, partly enclosed in the fuse holder or in an asbestos tube. The rating or capacity of the fuse is governed by the use of different gauges of fuse wire.
- *Cartridge*. These have a similar fuse holder or link, but the fuse element is contained in a cartridge of incombustible material filled with fine, arc-suppressing sand or a similar material. Once again, cartridges of various current rating are available.

In practice the fuse will be installed in a fuseboard or consumer unit, with the main isolating switch adjacent to the meter.

Portable apparatus may also be protected by using the standard rectangular pin plug, which will be found to have the cartridge on the live phase side. They are colour coded as 2, 5, and 10 A in black with 3 A in red and the highest capacity of 13 A in brown.

Miniature circuit breakers

These are an alternative to fuses and are used to protect circuits from excess current. These miniature circuit breakers (MCBs) are automatic switches which open when the current flowing through them exceeds the value for which they have been set. Variation of ratings can be obtained by making the operating mechanism operate at different currents. These operating mechanisms are usually an electromagnet and a bimetal strip. MCBs can be installed in a distribution board or consumer unit in a similar manner to fuses, and some boards are designed to accommodate the use of either.

Earthing

A fault or accidental damage may cause live conductors to come into contact with the metal casing of apparatus or other accessible parts of an installation. It is therefore necessary to ensure that should this happen the fault current does not flow long enough to cause damage or fire, and that there is no risk of anybody receiving a shock should they touch the metalwork. This is achieved by connecting any such metalwork to the general mass of earth, so that the resistance of the path to earth is low enough to ensure that sufficient fault current passes to operate the protective device. Equally, the fault current must take the earth path rather than the path through the person touching the metalwork. For this reason the impedance path of the earth must be low to ensure the path is efficient. The student, from Section 8.1 covering inspection and testing, will already be aware that he or she must extend the system checks, and to conduct an earth loop impedance, and a conductor-to-earth test.

Distribution fuseboards

An obvious method of connecting circuits would be to run a pair of conductors around a building and then to derive current from points where it is required. However, this should not be done. The supply ends of the various lighting and power circuits should be brought back to a convenient point in the building and connected to a distribution fuseboard, with each individual circuit protected by either a fuse or circuit breaker. For a domestic installation there will normally be only one distribution fuseboard or consumer unit. The supply is controlled by the consumer's main double-pole switch (often included in the consumer unit) so that the installation can be isolated when required. A pair of mains cables, of sufficient rating to carry the maximum current taken by the installation are connected from the electricity board service fuse and neutral link through the meter to the consumer's main switch.

In the larger installation the intruder alarm engineer will find several distribution fuseboards, each supplying one floor or section of the premises. In such cases the incoming mains are taken to a main distribution board, where they connect to a number of large fuses or circuit breakers protecting the outgoing circuits. Sub-mains cables connect the main board to a smaller branch distribution board or boards which will contain smaller fuses or MCBs protecting the actual lighting and power final circuits. The engineer who is to install the control panel and related equipment will have the alarm circuit connected at the main consumer unit in one of a number of ways; however, at the control panel/power supplies, he or she need only be concerned with the fitting of a fused spur.

Permanently connected appliances are normally connected through fused connection units, and other examples outside of the intruder alarm control panel and related equipment include wall-mounted heaters, extractor and cooling fans, hand driers and isolated lighting circuits.

The fused spur is available in many different forms, both insulated and metal clad, flush or surface mounted or with a cable outlet on the front for connection by a flexible cord. When these cable outlet fused connection units are used, the cord grip device must always be tightened after connection of the flexible cable cores. Flush-type fused connection units are used with standard boxes or pattresses. The engineer will find the fitting of a spur a relatively easy task once the testing and inspection of the mains supply has been done. He or she must, however, ensure that the control panel is connected to the out or load terminals with the supply from the consumer unit wired to the in side.

Testers are readily available for checking the mains supply. There are essentially two devices suitable for the insulation and continuity tests. The 'Megger' is a compact, self-contained, hand-operated type with a 500 V output. The test voltage is obtained from a hand-cranked, brushless AC generator, the output being rectified to give a DC voltage which is constant within a given range of cranking speeds. An alternative version features standard dry batteries and can therefore be operated by one hand. The test voltage is electronically generated, and only becomes available when a test button is pressed, hence it conserves the battery supply. The selector switch will normally be found to have three positions: battery check, megaohms and ohms. A specific tester is available to do earth loop impedance tests. It draws its power from the mains supply, and the values can be read directly in ohms. The instrument operates by passing a current of the order of 20 A for some 30–50 seconds from the phase conductor through the EFLI path via a known resistor in the tester. The voltage drop across the resistor is measured in the tester, from which the current and EFLI are obtained. The result in ohms can be read directly from the scale. Its operation is unaffected by supply voltage variations. These testers will generally include a recessed socket at the bottom for connection of the mains supply and a jack socket at the top for connecting test leads when testing at lighting outlets or at bonded metalwork. An automatic check of polarity and earth continuity can also be provided. If testing a socket outlet is to be done, it is only necessary to plug the device, via its flexible cable, into the socket. When testing at other points the tester can be plugged into a socket outlet and then connected by a flexible lead to the point concerned.

In so far as the mains supply is concerned, the intruder alarm engineer should be aware of Chapter 74 of the IEE Wiring Regulations, because these require completion certificates to include a recommendation that the inspection be tested at future intervals.

It is worth noting that cables in domestic premises changed from red and black to brown (phase) and blue (neutral) respectively in March 2004, so a mix of colours will be found for many years. These changes were confirmed at the time by the National Committee for Electrical Installations and published in Amendment No.2 to BS 7671: 2001 the IEE Wiring Regulations. This change aligned the electrical installation wiring colours with those of flexible cables of appliances and equipment.

Discussion points

One of the main occupational hazards that exists with respect to working on intruder systems is that they are ultimately powered by the mains electricity supply. We know that working safely with electricity is a fundamental requirement. It follows that the installer must:

- never work on a system connected to the mains supply unless recognized as competent under the Electricity at Work Act 1989;
- never remove labels from plant or machinery that he or she does not have responsibility for;
- never assume that a circuit is 'dead';
- never replace a fuse carrier that is simply found lying around;
- never leave a fuse carrier next to a fuse box even if the circuit has been labelled appropriately;
- never disconnect the earth wiring of appliances or circuits;
- never confuse the protective conductor, i.e. the mains earth, with any reference earth and always ensure that metal enclosures are correctly earthed.

A competent person is one who can demonstrate by acquisition of a recognized qualification, or experience or knowledge, that he or she can work safely on electrical supplies. A competent person also recognizes the limits of his or her expertise and will not undertake work that he or she is not trained for.

The removal of earthing conductors is extremely dangerous because the bond between all exposed metal in a premises and the final earth position ensures that an electrical path exists that will provide current leakage

Figure 8.2   *Illustration of reference earth to earth conductor*

protection in the event of a short circuit. The removal of an earth conductor can mean that part of the building can become 'live' with the potential to kill.

Reference earths provide a path for induced current to dissipate from, for example, the braid of a shielded cable. These reference points are deliberately taken to earth at one point to ensure that earth loops are not set up in the shield. However, the size of these reference wires is small compared to the size of a conductor that is needed to carry sufficient current to blow a fuse or operate an RCD, and it is important to understand the differentiation between them. This is clear by reference to Figure 8.2.

The tests that must be performed and recorded as part of the commissioning process under the Electricity at Work Act can be seen as:

● a visual inspection of the mechanical protection of cables and housings;
● an insulation resistance test;
● an earth loop impedance test;
● a polarity check.

We may conclude that safety is achieved by ensuring that the mechanical protection of the cables is intact and that the insulation between conductors and between the conductors and earth is within the requirements of the regulations. Also, the continuity of the earth conductor must not present a resistance in excess of that allowed by the regulations, and the conductors must be connected to the appropriate terminals with respect to polarity.

9 Commissioning, maintenance (servicing) and fault location

This chapter covers four areas. These are testing the system, regulatory authority requirements, fault finding and customer care. Consideration of each of these areas is important if a system is to be installed correctly. Ideally, the next time a technician should visit after installation is on the first inspection. If the system develops an early fault then the profit from the maintenance contract begins to diminish. If the fault is not cleared expediently then the system becomes both a technical and financial burden.

This chapter describes all of the required checks or tests and the typical results that are expected. These tests are vital to ensure that the system meets all of the requirements pertaining to it and is able to provide many years of successful operation. As a follow up to this there are many regulatory authority requirements to be satisfied. With this in mind the student must have a clear understanding of which authorities are to be notified on commissioning and the publications that are in existence to define the routine maintenance. Certain factors determine the periodicity of maintenance and the checks that must be performed, there are many details which need to be recorded.

This chapter covers the logical procedures for the location and rectification of faults and the use of test equipment to achieve these ends.

In the area dealing with fault finding it is important to be able to interpret circuit and block diagrams and identify the functions of components in a circuit. Equally it is necessary to be in a position to be able to deduce which component if at fault is liable to give trouble. Some knowledge of the effect on equipment performance of given component faults is required. The student should also know how to select suitable equipment and how to connect it and interpret the results for any given test.

There are also particular requirements for the resetting of remote signalling equipment in accordance with governing standards following faults, and diagnosis of the cause of a fault is vital in these circumstances. In certain cases there is a need to be able to diagnose which faults are likely to have been caused by electrical interference and the route by which this interference is entering the system.

Within the scope of customer care the installer must be able to identify the persons to whom it is appropriate to hand-over administration of the system and to identify the requirements of the users. There are also procedures to successfully instruct these users in the operation of the system and to prepare the hand-over documentation in clear terms. Alongside the hand-over documentation there is a need to advise of the maintenance programme to provide ongoing care of the intruder alarm system. Also, appropriate notification should be made to the relevant emergency services, and the installer should ensure that he or she has understood how this must be put into practice to prevent any communication breakdown in the event of an alarm activation.

We have come to understand that the intruder alarm industry is progressive and that a major change in this industry was the adoption of the EN50131 suite of standards. This involved a certain amount of parallel running with the traditional BS 4737. It also meant that the regulatory bodies needed to implement new checklists and completion certificates etc. in accordance with the EN. Due to a number of requirements needing to be ratified, a lot of the logging data forms still pay allegiance to BS 4737 but must be based upon compliance with PD6662.

9.1　Testing the system

Mechanical, visual, audible and environmental tests

Wiring should be inspected for compliance. Cables for intruder systems should be either plain or tinned annealed solid copper not less than $0.2\,\text{mm}^2$ in cross-sectional area, with a maximum resistance of $95\,\Omega/\text{km}$. A pro rata resistance is to apply to cables up to $0.5\,\text{mm}^2$ in cross-section. Stranded cables should not be less than $0.22\,\text{mm}^2$ in cross-section. Any strand joints should be brazed or hard soldered and not less than 300 mm apart. The tensile strength of these should not be less than 90 per cent of adjacent continuous cable. Cores are to be no fewer than seven strands of 0.2 mm conductor. Insulation should overlay the wire but not adhere to it and so prevent a clean strip. Sheaths should likewise not adhere to cores, and these are to be not less than 0.4 mm thick. Insulation resistance should be greater than $50\,\text{M}\Omega$ for 1 k at $20°\text{C}$.

The wiring should be inspected for any joints made outside of junction boxes or that feature unapproved methods of jointing.

Any damage to the cores of the wiring needs to be checked for, and also that there is no missing insulation or that it is stripped back too far. The cabling should be inspected to confirm that at no points has it been stressed.

In the visual tests, provision should be made to prove the consistency of colour coding. There must be correct segregation of the wiring from other services in the building. This must cater for screening of the wiring, and it must ensure that any suppression against interference is made. Wiring that has been placed in conduits needs to be inspected, and recorded in the log. The conduit should be checked for grounding and correct use made of conduit boxes and glands. The wiring should be inspected to prove that it corresponds with the plans and the claimed routes. It should also be checked for bridging loops.

The areas and points through which the wiring extends should be inspected, and confirmation made that the ambient temperature cannot go beyond that allowed. This should not normally be in excess of 30°C.

The mains supply should be inspected for use of an unswitched spur and the correct fuse rating.

The transformer within the control panel and any power supply should be inspected to prove that they are correctly fused and earthed, and that they are incapable of creating any hazard for materials or cables within the vicinity.

The power supply and control panel must be compared with the record to confirm that they are of the stated and correct capacity. Take note of the battery recharge time, the mechanical means of security afforded and that it is correctly connected, dated, and that it is not fitted within a sealed enclosure but is efficiently vented.

Audible tests are best conducted with the aid of a sound level meter under the worst ambient noise conditions. This test should apply to all sounders, internal and external, plus warning buzzers, which must be at the correct sound level at all prescribed exit/entry points.

Self-activating and self-contained bell modules must be additionally checked by removing the hold-off supply. It is possible to verify the bell cut off by disconnecting the load and ensuring that the signal times out correctly. The bell delay, if programmed, and rearm functions can also then be verified.

Operational tests

These must involve a full setting programme with checks against all the control panel functions, the detection devices responses and the correct operation of the signalling devices, both local and remote. The control panel must be checked to ensure that it recognizes the operation of every sensor. The sensor itself must sense the presence of intrusion consistently and repeatedly every time that it is challenged to activate. The sensor must also never false activate under any actual working condition in the protected premises.

The detection devices should be verified against both the manufacturer's data and also against the specification detailing the installation. Walk tests are easily performed using test LEDs and control panel audible test responses when checking movement detectors. There should be a level of tolerance to pulse counts since a number of events will be needed for full response.

All the tests should be carried out under normal conditions when there is ambient noise and heating in operation.

Anti-tamper loops can be broken to prove the correct response in set and unset conditions, and this can also extend to junction boxes and lid tamper protection.

Door contacts should not activate unless the correct level of movement is achieved, and vibration detectors or glass break devices should have their sensitivity adjusted to suit the pertaining conditions.

Electrical tests

The system installer must provide and retain records of the installation. Electrical test results should be on system log book test sheets.

These tests must cover:

- *Isolation*. The circuit insulation resistance, ground loop impedance and continuity from the mains to the panel and/or power supply unit need to be tested. The grounding and isolation of transformers in use and the voltage at the primary side should be checked.
- *Power supply*. The output voltage and current for the high- and low-charge voltage need to be checked. The separate battery outputs should be measured and the battery proved to be fully charged. The battery resistance should be recorded with the system in a set condition and with the mains supply disconnected at source. The polarity should be correct and the neutral not fused.
- *Sensor detection loops*. It is necessary to measure voltage, and supply current under both alarm and quiescent conditions plus resistances. This is achieved by placing a low-reading ohmmeter across each detection loop and paying regard to the number of detectors on each loop and the internal resistance quoted by the manufacturer of the device. Even a reed contact resistance can range from 100 to 250 mΩ, and powered devices will have an in-line resistor. All circuit resistance should be recorded. The reading should be infinity across the anti-tamper loop and the detection circuit. The resistance between the earth point and the detection and anti-tamper loops should also be infinity.
- *Alarm demand current*. To calculate the required current in an alarm condition, add all the current consumptions of all detectors, sounders, strobes, communicators and any other devices wired to the system. This will provide the total alarm demand current.

Subtract the alarm demand current from the power supply output current. Any extra current requirement will need to be supplied by the battery in an alarm condition. Note that if the battery is not capable of supplying the extra current requirements in an alarm condition for

Control unit –				Contract No. –			
Subscriber –							
Address –							
Installation engineer –							

Zone	Description	Detection device	Voltage at detector	Current (mA) at detector	Resistance Ω		
					+	Loop	–
1	Kitchen	PIR (320)	13.65	14.10	31.2		1.2
2	Lounge						
3							

Power supply	Location	Battery AH	Normal V. out	Battery (Mains off)	
–	–	3.0	13.65	I. load	I. alarm
				V.out	

Figure 9.1 *Essential readings*

the period of the alarm, it will be necessary to increase the battery size. Ensure that the automatic switching of the system from mains to battery functions correctly.

Communication tests

- Confirm that the dialling equipment is securely positioned within the protected area.
- Confirm with the alarm receiving centre the message that is to be generated, and have the ARC confirm its receipt.
- Produce documents to confirm the level and type of responses that have been agreed.
- Recognize the essential readings that should be logged for future reference (Figure 9.1).

9.2 Regulatory authority requirements

In Chapter 1 we gave an overview of the response organizations that are active in the intruder alarm industry together with the inspectorate bodies. It is now a convenient time to look at the alarm system and how the police will view any response to it, as many installers believe that should their alarm activate the police will respond immediately.

Under the ACPO Security Systems Policy 2000, the police will only respond to alarms that have been credited with a URN. Installation companies must therefore be within approved schemes or the customer cannot expect a police commitment to respond to an alarm condition. In addition the system must be installed, maintained and operated in accordance with the ENs. Therefore a maintenance contract is essential as without one the customer is deficient.

In the ACPO policy there are two signalling types.

- *Type A.* These cover alarms which transmit a signal to a BS 5979 recognized alarm receiving centre or to a specific police control room.
- *Type B.* These are audible only but include speech dialler systems.

In regard to Type B the police are only committed to respond if a third party can provide confirmation that an offence is in progress. A neighbour must see or gain evidence that an intrusion is actually happening. Therefore it follows that the way forward for the intruder alarm engineer is to have systems linked to ARCs as these can also carry out false alarm filtering so that only genuine transmissions are notified. Systems fitted with new generation speech diallers are gaining wide popularity and have a definite role to play, but these must never be programmed to dial police numbers in any direct way.

The policy stipulates three levels of response:

- *Level 1 response.* This is immediate. It means that the police will respond to an alarm with no delay.
- *Level 2 response.* Although an immediate police response is desirable, such action may be delayed if there are other situations which take priority, such other duties or responses being required against level 1 alarm activations.
- *Level 3.* This is keyholder only.

When a system has been credited with a URN it automatically qualifies for level 1, but the police response will change if the system creates false alarms. The extent of these changes is detailed and shown in a tabulated form in Section 1.5.

It follows that in order to obtain a realistic police response the quality of the installation and compliance with the necessary criteria must not be open to question. Connection of the system to an approved central station remains paramount. The other consideration is that of maintaining level 1 response, and that is achieved by confirming the nature of any alarm activation. This brings us back to confirmation technologies, which are endorsed by the ACPO policy to include visual, audible and sequential confirmation. They are a recognized and approved method of reducing false alarms and also receiving the desired police response.

There will always be companies who offer alternatives to what is laid down in the ACPO policy, such as response from a private security firm or non-approved alarm signal monitoring. These firms cannot be regulated or licensed, and in practice are not ideal. Private security firms can be almost anyone, perhaps including criminal elements. These points must be borne in mind by the intruder alarm engineer when representing his or her company and the service of quality that is offered with support by URNs. Some installation companies may attempt to compete by claiming to offer certificates of compliance, but these will be worthless when it comes to offering URNs. It is always a question of educating the customer regarding qualification for police response and how this is governed by the installer selected.

9.3　Fault finding

Successful and efficient fault finding or troubleshooting depends on logical reasoning and organization. At times it also depends on luck, but the more reasoning that the engineer is able to use, the less good fortune he or she will need. It is vital to understand fully the circuits in which the installer is interested and how they are engineered to function.

How circuits are actually functioning compared with how they should function is the first basic information that is needed.

We can often be guided by the control panel log to detection devices that appear to be faulty or at least the detection circuit in question. However, due to the different types of circuit that may be employed, and the fact that line monitoring devices can be used, the engineer must understand the fundamentals of certain electronic components and how they operate. Without some knowledge of these he or she can never be fully aware of the effects of given component faults on the performance of equipment. With this in mind we shall look at the most common components that the intruder alarm engineer will encounter.

Resistors

These are probably the most common electronic component, and are designed to introduce a specific amount of resistance into a circuit. They are available in many sizes and ratings. As a resistor works it becomes warm as electrical energy is converted into thermal energy. If the resistor becomes too hot it will be damaged, effectively changing its resistance properties. It is important to be aware that the term 'larger', when applied to resistors, refers to its wattage rating and not to its physical size or resistance value, which is measured in ohms (Ω). It is worth remembering that if any doubt exists, a 'larger' resistor should be selected, choosing,

Figure showing Resistor colour code chart.

Resistor colour code

4-Band code   5 6 000 ±5%   56K-ohm ±5%

	1st Digit	2nd Digit	3rd Digit	Multiplier	Tolerance
Black	0	0	0	1	
Brown	1	1	1	10	1%
Red	2	2	2	100	2%
Orange	3	3	3	1000	
Yellow	4	4	4	10000	
Green	5	5	5	100000	
Blue	6	6	6	1000000	
Purple	7	7	7		
Grey	8	8	8	0.1 Gold	5% Gold
White	9	9	9	0.01 Silver	10% Silver

5-Band code   47K-ohm ±1%

4 7 0 00 ±1%

Note: On all of the colour-coded resistors, the band at one end will be spaced further apart than the others; the resistor should be viewed with this band to the right to correspond with the chart and examples.

4-band codes

Reading from the left, bands 1 & 2 are the significant digits [1st green = 5, 2nd blue = 6].
Band 3 is the multiplier [orange = ×1000].
Therefore the value of our example is 56 × 1000 ohms or 56K.
The 4th band indicates the tolerance [gold = ±5%].

5-band codes

Reading from the left, bands 1, 2 & 3 are the significant digits [1st yellow = 4, 2nd violet = 7, 3rd black = 0].
Band 4 is the multiplier [red = ×100].
Therefore the value of our example is 470 × 100 ohms or 47K.
The 5th band indicates the tolerance [brown = ±1%].

Figure 9.2   *Resistor coding*

an example, a $\frac{1}{2}$ W resistor over a $\frac{1}{4}$ W resistor. In appearance resistors are typically cylindrical, and will have either four or five coloured bands that identify the resistance value. The colour code is explained in Figure 9.2.

Fixed resistors

These are so called because they have resistance values that do not change within a certain tolerance (unless damaged). Although all resistors exhibit some resistance changes in response to temperature fluctuations, these are insignificant and can generally be disregarded. Common types of fixed resistors are 'composition' and 'metal film' resistors.

Composition resistors consist of a thin coating of carbon on a ceramic tube. Because carbon conducts poorly, this type of resistor, although small, can create large resistances.

Metal film resistors use a thin metallic film instead of carbon and can be manufactured to exhibit more clearly defined and precise values. They are also less sensitive to temperature fluctuations than their carbon counterparts.

Variable resistors

Variable resistors, which may also be called rheostats or potentiometers, are used when it is necessary to change the resistance in a circuit. In practical terms rheostats and potentiometers are interchangeable, although rheostats tend to be used for heavy duty AC applications, with potentiometers being found in low-power circuits. Potentiometers tend to have three terminals while rheostats only have two.

Combining resistors

It is always possible to increase or decrease resistance by using a number of resistors in conjunction with one another. If the resistors are connected in series the sum is additive, i.e.

$$R_t = R_1 + R_2$$

where R_t is the total resistance. Hence, $1000\,\Omega$ can be obtained by connecting two $500\,\Omega$ components in series since the current must flow through both.

However, if two resistors of equal value are connected in parallel, equal current must flow through both. If the two resistors are not equal, then more current must flow through the path with less resistance, and the formula becomes

$$R_t = \frac{(R_1 \times R_2)}{(R_1 + R_2)}$$

The formula for several resistors connected in parallel is

$$R_t = \frac{1}{1/R_1 + 1/R_2 + \cdots + 1/R_n}$$

where n is the total number of resistors.

Figure 9.3   *LED 12 V DC circuit*

Diodes

A device that only allows current to flow in one direction is called a diode. Semiconductor diodes will break down at certain voltage levels, and the specification for a diode will list its peak inverse voltage (PIV) or peak reverse voltage (PRV) – the point at which it will break down.

Zener diode

These diodes function differently from regular diodes, in the way that they respond to reverse voltages. The voltage rating of a zener diode is the point at which it begins to conduct when reverse biased, and may be called the 'avalanche point'.

Light-emitting diodes (LEDs)

Similarly to other diodes these conduct in one direction only. As its name implies the device emits light when forward biased. The most common colours for LEDs are red, green and yellow (amber), but other colours such as blue and purple can be sourced together with some components that emit light in the infrared region.

The LED is essentially a durable component intended for low-voltage circuits, and most operate at 3–6 V DC. Any LED connected direct to a 12 V DC power supply will be destroyed, but it is possible to incorporate a 1 kΩ $\frac{1}{4}$ W resistor in line to operate as an indicator (Figure 9.3).

Since LEDs are forward biased and emit light only when polarity is observed, they can be used to check polarity.

Capacitors

Voltage is stored in capacitors. It is the insulating material – the dielectric – in a capacitor which helps determine the capacitance, and a capacitor is generally classified by the type of dielectric it contains. The capacitor can therefore be seen as a reservoir in that it stores voltage until a threshold is reached when it must then release the voltage.

The unit of capacitance is the farad (F), but in most circuits it is too large to be practical so the microfarad (μF), which is equal to one millionth of a farad, is used. Most capacitors are rated at two voltage levels: breakdown and working voltage. The breakdown voltage is the maximum that the dielectric can withstand before breaking down. The working voltage is the maximum DC voltage that can be placed across the capacitor plates safely. Exceeding the working voltage may damage or destroy the capacitor and other components.

In summary, the capacitor is a passive electronic circuit component usually consisting of two metal electrodes or plates separated by a dielectric (insulator). The capacitance is the property exhibited by the conductors separated by the dielectric, whereby an electric charge becomes stored between the conductors.

With some knowledge of the function of the main electronic components we can better understand how faults and false alarms are generated. However, units of equipment or devices do fail, so in order to cure problems certain systems do at times need main items replaced.

Nevertheless, in the first instance we must take a proper attitude towards reliability and the consequences of unreliability, that is, the need for maintenance and service. It is vital that the manufacturer designs for reliability, and the installer must install to the same end. Practically all intruder detection devices and control and signalling equipment incorporate semiconductors, diodes, transistors and integrated circuits which are reliable in principle yet are manufactured in vast quantities. These components when given their rated value are also given a tolerance. Accepting tolerances means that the unit cost can then be lower, but the designer must ensure that no matter what combination of tolerances exist in the equipment the final product must still function as intended at all times. The designer is hence committed to design not only for initial manufacture but also for future maintainability and serviceability. In addition, designs must change and expand as new functions arrive on the scene. An extremely reliable control panel with few functions to compete with new-generation equipment would clearly not be able to command a reasonable share of the market. For this reason devices must be constantly updated, and this of itself generates certain problems. Fortunately, with computer-aided design, changes needed to allow for tolerances can be calculated and assessed more quickly so that substantial improvements in electronic equipment and reliability are gained. However, components can still fail, and this, allied to the unfeasibility of checking every component, will always mean that some uncertainty must remain. Nevertheless, the overwhelming evidence is that electronic security equipment is essentially inherently reliable.

The intruder alarm engineer will always use his or her experience in the selection of equipment, and must ensure the correct and efficient installation and maintenance of the system.

Wiring

One immediate step that the installer can take in a system that is at fault is to examine the normal interface between the equipment and the wiring. This can refer to wiring connectors, terminal blocks, or any other means by which wires are joined or terminated. It is at these interfaces that so many installation and service problems arise.

Copper is the normal material for interconnecting wire, and perhaps we should have some understanding of its nature. As supplied in wires and cables it is 'half hard', meaning that it is neither too hard nor too soft. In this condition, however, it is easy to bend yet remains incredibly strong. When bending copper, the inner portion becomes compressed whilst the outer is stretched, which makes it slightly 'work harden'. The wire in consequence loses some of its flexibility. Another problem concerns terminal blocks featuring pinch screws for securing the wire in the block, as the screwing down of the pinch screw on to the wire compresses the copper over a small concentrated area and makes it more brittle. Provided that the wire is not disturbed from then on, the connection can survive reliably, but in routine maintenance and in fault finding there may be a need to remove it. This can cause fracture of adjacent wires that have become brittle, and the fault may not become apparent until a later stage. In practice the thinner the strand of wire the less the work hardening when the wire is bent but the lower its mechanical strength and current-carrying capacity. Single-strand wires certainly have a poor history of failure through fracture in service. The multistrand cable, using multiple strands of thin wire, is far more reliable, and tends to be used in the intruder industry with the strands twisted together. The use of multistrand cable is further improved in reliability when used in conjunction with terminal blocks using flexible strips between the wire and screw. Wires are not damaged when the screw rotates although it is a little more difficult to achieve the same level of pressure on the wire compared with single-strand cable. If great care has been taken with the connecting of detection loops and other wiring then there should be no reason to suspect bad connections are the cause of faults unless another person has worked on the system at some other point and not applied the same care. It is also possible that cables have been damaged in some other way by other trades, and these faults may not appear until some later stage.

Stages in fault finding

There are many ways that a fault can be generated in a security system because of its inherent sensitivity, and it will be of benefit to commence fault finding by following a set pattern:

- Talk to the end users. Obtain information from them, but keep an open mind.

- Check the symptoms.
- Observe the status lights on the control panel.
- Interpret the LED or LCD displays for information and enter the engineer's log for events. This may simply define a component or module as necessary for replacement.

It can be useful to view the control panel as a four-block function as in Figure 9.4.

In the first instance we can look at the power supply. In Chapter 8 we have studied how to inspect and check the mains supply. We can now assume that this has been done, but should confirm certain factors:

- Verify that the mains supply is correct and that no disconnection has been made.
- Confirm the low-voltage AC secondary side of the transformer is in accordance with that quoted for the equipment.
- The AC input should be within 5 per cent of the intended supply on the primary side between the phase (live) to neutral, and the phase to earth should be the same.
- Using a low AC reading to ensure that the reading between neutral to earth is not greater than 0.5 V AC. Establish that at the time the fault occurred there was no power cut in the protected premises.
- Confirm that the auxiliary supply power is of the order of 12–13.8 V DC and in line with the original voltage at the last inspection.
- The battery charge input/output must be confirmed. The DC charging voltage should read 13.5–13.8 V DC.
- The signalling processing board/user interface should be stabilized at about 5 V DC.

Having checked the power supply it is then appropriate to look at the programming and user interfaces.

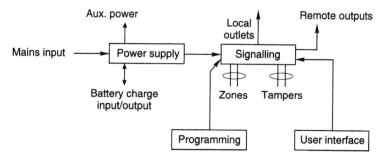

Figure 9.4 *Fault finding*

The information at this point will be very much governed by the logs that are available and the information that the end user can provide. It is possible that the engineer can be guided by such information, but he or she must also consider that any information given by the user may be incorrect. Only logic can prevail at this point, and the engineer must guarantee that the fault was not purely operator error as a result of new personnel. Proof should also be gained that the system use has not been changed or corrupted and that a change in programming is not now required. We can then look at the signalling, starting with the zones.

A walk test should be performed to check all detection devices for correct operation. Use the setting and unsetting procedure and observe the status lights on the control panel, as this may establish a detector at fault. It is possible that at this point a detector is shown to be at fault, but checks must extend to prove that there has been no change in the area that the detectors are installed that could influence its operation. There are false alarm hazards for all detector types, but has some recent change in the premises influenced the detection devices?

Visual checks against any detection device found to be in a zone causing faults should be performed.

The engineer must then move his attentions to the wiring of the zones.

The first thing to consider is what type of wiring system is being employed. This may be a conventional, using four-core cable, or there may be end-of-line devices employed. Knowledge must also extend to the number of detectors used on the zone wiring to establish what readings are expected at the control panel terminals once the detection loop has been disconnected from them. With a conventional wiring system the alarm loop resistance should be low and should be verified against the original installation readings as it will depend on the number of detectors employed and whether they are active, and influenced by the integral loop series resistor, or if they are non-active simple contacts. With non-active contacts the tamper resistance should be low and of the same order as that of the alarm loop. Between the alarm and tamper loops the reading should be infinity (open circuit). In the event that end-of-line resistors are used, the readings should be considered against the manufacturer's data and original readings. With active detectors the loop series resistor, which will be in the region of $10–50\,\Omega$ depending on detector type, needs to be taken into account. Open circuits will show no continuity and are easily found, although active detectors must be powered to be closed.

Simple continuity checks can establish breaks in wires and shorts by the disconnecting of the cables at both ends.

Direct-reading capacitance meters can be used to measure the capacitance from each wire in a cable to earth. High capacitance readings will indicate wires that are intact whereas low capacitance readings will point to broken wires. If a ratio between two values is calculated it will also give a ratio of the distance between the control panel to a break and that of the full run, so an idea is given of the area where the problem exists.

With the circuits intact the polarity of active detectors can be checked and the current drain of the load measured, which must be of the order of 10–40 mA 12 V DC.

The control signal for first to latch detectors must also be verified.

Figure 9.5 shows how we can consider using the latch freeze facility to identify detector or wiring faults (see also Figures 5.1 and 5.2).

Checks can now be extended from zones and tampers to the outputs on sounders and the signalling equipment.

The correct polarity to stand-alone bell (SAB) or self-contained bell (SCB) modules should be verified, and also the required voltage at the module. The tamper should be confirmed as positive- or negative-return single wire. If a dedicated 12 V supply is provided from the hold-off and trigger points at the control panel to the sounder unit and strobe, these devices should operate. If not, the bell box is at fault. Speakers can be checked by pulsing them.

The bell box standby can be checked by removing the hold-off voltage at the control panel.

Because of the many combinations of wiring control panels and sounders through SAB or SCB modules and the vast array of products on the market, the checks to prove that they are fault-free are limited to the above.

Remote outputs are limited to checks against supply voltage, correct wiring of the line and the trigger signal to the communication equipment from the control panel.

Figure 9.5 *Set output*

Interference

Many of the faults that we have studied and the checks that can be made to confirm them are easily understood because they can often be seen in practical terms. A range of problems, however, are environmental, in the form of various types of electrical interference, conducted, induced or radiated. In Section 5.6 we have looked at voltage surges and induced electromagnetic energy, but to what extent does this appear as a fault or a source of false alarms?

In the first instance we know that increasingly sophisticated control and signalling equipment is being more widely used, and this even extends to the most basic domestic alarm system. We also know that this increases complexity and thus their vulnerability to environmental problems.

False alarms tend to be caused by the following although it is difficult to establish exactly to what degree each is guilty:

- customer misuse
- installation fault
- equipment fault
- unidentified external effect.

The first three causes of false alarms are quite easily defined but the problems caused by unidentified external effects such as Radio frequency Interference (RFI) and electromagnetic interference (EMI) are rather more complex.

RFI

This is largely the result of radiated signals in the atmosphere close to alarm installations. The causes are:

- mobile radios in police cars, fire engines, taxis, etc.;
- close proximity to low-flying aircraft;
- military aeroplanes with powerful radar equipment;
- illegally boosted CB radios (cellphones and legal CBs do not cause problems).

The protection and the suppression of the problem is mainly the responsibility of the intruder alarm control panel and the detection equipment designers and manufacturers. This must cover aspects of design, including levels of RFI protection. Adequate filtering and buffering are needed.

EMI

Mains-borne interference is well known to those using personal computers and other microprocessor-based equipment that has 'crashed' when electrical equipment has been switched on or off. Indeed, it is recognized that even in the domestic environment the mains supply can be 'dirty' in that it is contaminated by numerous voltage surges and high-voltage transient spikes.

Spikes are conducted or radiated and enter the equipment via the mains supply. A radiated spike is induced into the cable supplying the unit when the cable is of a critical length that relates to a transmitted signal wavelength. It becomes a very efficient half-wave- or quarter-wavelength aerial.

Although all electrical equipment can cause a spike when switched, fluorescent lights, air conditioning units, oil-fired boilers and devices containing motors can induce large voltage transients. Lightning striking nearby, supply transformers or overhead cables will cause the same phenomena. EMI can cause spurious actuation of an alarm, and in the case of a microprocessor a pulse can be read as an instruction leading to a system crash or erasure of a memory database.

The answer is to resite the mains supply and install a filter. Filters in the form of a fused spur are advocated. These reduce the chance of false alarms from mains spikes and RFI by using effective radiofrequency filters and transient voltage suppressors, by means of inductors and capacitors, along with varistors to suppress transients.

Microprocessor-controlled equipment needs a stable 5 V DC supply free from surges, spikes, voltage fluctuations and radiated electrical noise. The regulated supply to detectors must also be free from interference. Filters are designed to comply with BS 1362 and BS 1363 and be correctly earthed. Figures 9.6 and 9.7 indicate the cumulative result when tested to BS 613.

An important part of this chapter is to inform the student how to diagnose which problems are likely to have been caused by electrical interference and to establish the route by which the interference is entering the system. It is therefore appropriate to study the subject again in general terms with special regard to conducted, induced and radiated interference.

Conducted interference

This is mostly generated by the various forms of high-speed electronic switching of power equipment connected to the mains electrical system. When the mains supply is used to maintain battery supplies, which it must for security systems, the mains-borne interference finds its way into the equipment and results in unwanted and erratic switching of its functions. Such interference should be suppressed at source, and therefore suppression must be considered at the design and manufacturing stage.

Induced interference

This tends to be continuous and at a level too low to cause unwanted switching yet high enough to interfere with wanted signals. It originates in the mains power supplies and from cables lying too close and parallel to signal cables.

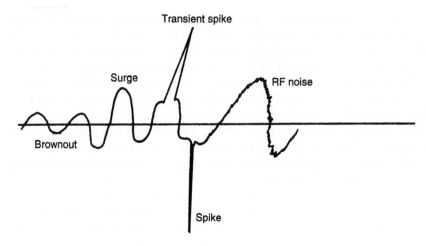

Figure 9.6   *Mains voltage variations*

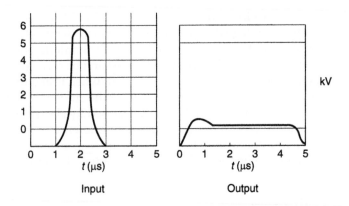

Figure 9.7   *Spike rejection*

Radiated interference

The most extreme form of radiated interference comes from lightning, and because of the very high energy involved there is also some measure of induced interference. There is always a high incidence of false alarms coinciding with electric storms and particularly at short distances from the storm centres. For fringe systems, however, the methods adopted for limiting the effects of switching spike voltages on power supply mains inputs to equipment can be used to assist in the reduction of the interference attributed to electric storms.

Radiated interference is also attributed to radio transmitters such as police radios and CBs in vehicles. Fluorescent lighting also falls into this category, and its interference can be treated with low-inductance flat capacitors

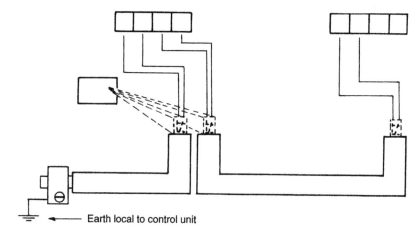

Earth local to control unit

Figure 9.8 *Earthing single point*

connected close to the component most susceptible to the interference and a ground point. If the interference is being picked up by the component itself, such as an inductor, the component needs to be screened with a metallic cover also connected to a grounded point. Earthing itself is of major importance, and is thus considered separately.

Earthing

In semiconductor electronics only low values of current are generated, and because short lengths of leads or printed circuit board tracks are involved only negligible voltage drops occur. However, when we consider interference signals, the current, and hence the voltage drop in earth leads, can be so much higher than normal equipment working levels. To avoid any malfunctioning from interfering signals it is a help if all the leads to ground are taken to a single point, including the power supply. Therefore, earthing is usually at the power supply, and any screening used to protect cables from interference is also connected to that point. To maintain single-point earthing the screen has to be sheathed in an insulating sleeve to stop the screen contacting another earthed medium, such as a fence, and forming an earth loop. This would of course affect the properties that are required of the screening and single-point earthing. It is apparent how the screened part of screened multicore cable should be run and earthed at a single point by reference to Figure 9.8.

We can conclude that we should at times be guided to the presence of faults by the control and processing equipment logs; and printers may be adopted as ancillary equipment as event recorders to hard copy system information. Indeed, high-technology control panels can provide values for current consumption, battery charging voltage and circuit resistance and imbalances. This all helps to trace problems. The question of interference, however, is more far reaching, and must be appreciated as a complex problem. It has been given varying levels of scrutiny within appropriate chapters in this book depending on the subject at the time, but must be understood as a whole.

The subject of fault finding cannot be concluded without confirming how it leads to false alarms and how this has an effect on remote resetting.

The ACPO policy reinforces engineer reset for remote signalling that has police response. The system may be reset and rearmed remotely by the central station if adequate information is available for the maintenance engineer to resolve any false alarm problem. This is to ensure that important information cannot be lost if the customer resets. This would certainly lead to additional false alarms. The system installer must ensure that the customer is fully aware of the use and function of the alarm system.

Remote reset is a function that can be performed with particular Red-CARE transmitters and interface boards. Following the normal sequence of events the end user calls the central station, and a verification procedure is followed. A 'tellback' signal will be sent to the control panel through the RedCARE network, and the user can enter a valid access code to reset the system. Interfaces allow the selection of relays such as PA (personal attack) and intruder to be normally energized or de-energized with relay states selected by links.

Remote reset has been considered in more detail in Section 5.2, to which reference should be made.

9.4   Customer care: servicing and maintenance

At this stage the installation company will be at the point of handing over the intruder alarm to the client, and must understand that in addition to presenting the system and related paperwork in a professional fashion the installer must also brief the customer on the need for servicing and maintenance.

It is accepted that the inspection, testing, commissioning and setting of the system in question must be exhaustive. The hand-over must also follow this philosophy. All these factors are absolutely vital to reduce false alarms and to heighten client satisfaction. Nevertheless, the client must also accept a measure of responsibility for the intruder alarm and using it in the correct manner. To this end the client should be furnished with a test certificate plus the related drawings and a record for the installation. The client must also be given a full demonstration of the operation of the system and be made aware of any duties that must be performed that are necessary for the efficient running of the system.

The installer must:

- Ensure that the intruder alarm system is exactly that which had been detailed and originally specified.
- Prove that the level of workmanship has been employed to the required standard and that the system has been proved to operate correctly. A record should have been established for its full testing and commissioning.
- Confirm the effectiveness of the system, paying attention to those factors enhancing reliability.
- Accept a measure of responsibility for ensuring simplicity of use of the system.
- Furnish the client with a manual for operation based on the data provided by the manufacturer of the control panel. This should give a general system description with information on the detection devices that are employed in the protected premises and the zone descriptions that apply and the locations.
- Ensure that the operating manual is precise with respect to the daily setting and unsetting procedures. Alarm call-outs and the method of obtaining a response from the system installation engineers must also be detailed in the manual.
- Ensure that telephone numbers are entered in the manual to cover any general requirements.
- Provide demonstrations of the system. These are of vital importance in order to ensure that the client has confidence in how to operate the system and the functions available. The installer must demonstrate full system operation and provide some degree of technical detail of the technology that the sensors use in their detection process. The engineer, however, must beware of confusing the client with too much technical detail. At the end of the demonstrations the installer should ensure that the client is confident in using the system as intended and satisfied as to its capability.
- Instruct the client not to position objects in front of movement detectors and to inform the installation company if there is to be any development at the protected premises in case this affects the system capability.

Following the demonstration of the system to the client, the records should be formulated for future reference. The criteria for records are as follows:

- *System.* Detail the specification, provide drawings and record how the maintenance and instruction manual was prepared. Make a note of any other points that could be used for future reference and any features particular to the given installation.
- *Historical.* Log the date and time of completion. It is important from then on to keep a record of every visit to the premises and the exact time. These

records should also account for any other engineer calling on behalf of the security company, and the name of that person must be logged. The record must include a signature by the person making the visit and any work performed. The nature of any fault must be recorded with a diagnosis, action taken and the time that the work was concluded.

- *Preventative maintenance.* This is the historical data plus the detail of the work performed and any faults that have been found. This is to include operation by the client and any faults generated by the client together with faults attributed to the equipment.
- *Corrective maintenance.* This is the historical and preventative data plus a description of the reason for the call-out together with the fault found and how it occurred. A log of the action taken must be made.
- *Temporary disconnection.* This to account for preceding data with any information regarding the details of disconnections made. The reasons for disconnections are to be recorded with the dates and times of replacements and resets.

The ACPO policy of course reinforces the need for servicing and maintenance. The maintenance that should apply is governed by particular preventative and corrective maintenance measures. The essential difference between preventative and corrective maintenance is as follows:

- *Preventative.* The routine servicing of a system carried out on a scheduled basis.
- *Corrective.* The emergency servicing in response to the development of a fault.

All alarm systems are essentially sensitive, and for that reason need comprehensive and regular maintenance. This to the end user may appear somewhat expensive, and because of this will want justification and positive results. Investigations must therefore be implemented to identify potential faults or false alarms. Poor attention to preventative maintenance will result in alienation by the client who can only conclude that there has been no reward for the financial outlay.

In the first instance the system would have been designed with a specific life span. Thus, the mean time for the expected failure of certain components (plus a safety margin) should have been taken into account. This factor influences the overall system price and quality, and it is from this that the consequential maintenance frequency and costs arise.

Preventative maintenance should be carried out within 12 months of commissioning on all local, audible-only mains supply and rechargeable battery standby power systems. Maintenance should also be carried out within 6 months of commissioning on all local audible or remote signalling systems using primary-only or rechargeable battery systems. Therefore, for preventative maintenance, for bells-only local signalling systems, in order to satisfy industry practices, maintenance needs to be performed annually, but for remote signalling it must be twice a year. However, sections of codes of practice that may be implemented by an approvals council may impose other stipulations in addition to imposing minimum standards of maintenance that should prevail. These standards must be complied with by the contractors.

BSIA Form No. 177 'Commissioning and preventative maintenance of intruder alarm systems installed to PD6662: 2004 Code of Practice' stipulates the following for the frequency of preventative maintenance visits.

Grades 1 and 2X	**One site visit per year**
Grades 2 and 3	**Either two site visits per year**
	or
	One site visit and one remote system check per year
Grade 4	**Two site visits per year**

Note: 2X is non communicating
Note: The interval between preventative maintenance should not exceed seven months (thirteen months for Grades 1 and 2X)

Even with the best systems in existence, regular maintenance and emergency maintenance may always be needed as a result of accidental damage or indeed vandalism or intrusion. Although most systems are now purchased outright, the client should always be offered a choice of maintenance agreements in case the user elects to rent or lease the equipment. In the case of the installation cost being high the client may be better advised to rent or lease. If the equipment costs less and is installed in a more difficult environment the maintenance contract may offer comprehensive cover. In such situations the probability of problems occurring may be greater, and service and maintenance could then include, as an example:

- full maintenance twice per annum;
- a clearly defined number of 'no charge' normal working hour calls per annum;
- a repair service for emergency situations outside working hours and chargeable at an agreed rate.

In the event that the client leases the equipment, then the cost for any components or materials used to effect a repair is not chargeable to the customer.

If it is predicted that the risk of trouble is lower and the equipment is of an extremely high quality, then minimum cover at a reduced cost may well be adequate. This type of cover may comprise, for example:

- full maintenance every 6 months or less in accordance with the approval authority stipulations but with the labour and materials required for any repair provided at no cost to the client;
- all other repairs at any other hours to be chargeable.

We must remember that maintenance includes both testing and inspection. Also, maintenance can only be effective if the installation company employee is trustworthy and trained correctly.

There are grades of service for emergency call-out corrective maintenance programmes. A level of service should be established that both the client and installer can agree to, taking into account:

- the maximum period of time a system may be out of commission;
- the maximum period any component of the system can be out of commission;
- the worst possible failure case.

With this in mind the client should be able to determine:

- the time needed to contact the system repairer;
- the time needed for the system repairer to travel to the premises;
- an estimated time for repairs to fix a given component;
- the time needed to recommission, test and complete each repair;
- the provisions that must be made for on-site or repair company stock replacement;
- which components are quicker and more cost-effective to repair in preference to replacing;
- any costs for overtime and additional guarding that would come into effect as a result of a failure to respond.

A document can then be drawn up to detail the level of service arrived at. The agreement could state the following for example:

System agreement. Attend site within 2 hours of receiving an authorized instruction. Provide all spares by holding stock of all replacement items x, y and z. Complete repairs under the worst failure case within 3 hours.

Of course, the level of service must be clearly defined, otherwise compensation may be sought by the client in the case of default by the alarm company.

With regard to corrective maintenance, this must be applied to all conditions be it emergency responses to a client, or alarm receiving centre calls to correct either faults or false alarms.

The response time, which is defined as the amount of time that elapses from when a call is received to when the engineer is on site, should not exceed 4 hours. An exception may be made for an offshore island site or a local audible bells-only system. If this 4 hour period is impractical or the client does not require it in order to reduce costs, then, with the insurer's agreement, it can be extended to a longer term.

It is always important from a customer care viewpoint to present certificates of training and competence levels of the staff who will be responsible for work on the security system, and these should be kept at the site and held by the security staff and keyholders. The preventative maintenance frequency has already been considered, but the tasks are to include as a minimum:

- Checks to cover the installation and location/siting of all equipment against the system record and the satisfactory operation of each sensor, detector and processor. This must be more than a simple panel function test.
- All flexible connections must be inspected, also all power supplies and control equipment.
- A full audibility and functional check must be performed. Attention must be focused on the performance of all audible and other signalling equipment.
- Control panel function tests must also be conducted.
- A log must then be made of the system of records. This must be protected from unauthorized access.
- The historical records must include dates of any visits to the protected premises with the faults being noted and the remedial action taken being fully documented.
- All alarm calls must be logged, the details of the action taken and, if known, the cause.

In Section 9.3 dealing with fault finding we looked at the procedures to be followed in the event of false alarms. These initially range from checking the subscriber's operational procedure to ensuring that the premises have not undergone a change in use or structural alterations. Changes in electrical supplies must also be checked for. The environment must also be checked to ensure that there are no changes that could create a possible source of environmental interference. We should include within this notes of any new heating systems, automatic lighting controls or possible introduction of airborne interference. These checks can all be made in the company of the client.

The customer also of course has a huge role to play in keeping the credibility of the system high. Not only must the engineer exercise care during installation, servicing and follow-up procedures, but the client must understand that many false alarms occur as a result of operator error. Most of these unwanted activations occur when the premises are being locked up for the night or opened in the morning. Whether it is the result of checking in a lock for a correct key, going back into a protected area after the alarm has been set, leaving a door or window open, or inadvertently pressing a personal attack button, the false alarm problem is the same.

The skill in implementing security therefore extends beyond the detection and signalling of deliberately operated devices to the avoiding of false alarms. With the client creating the problem, once the engineer has ensured that the system is simple to use then tact is required. Nevertheless, we should avoid attaching blame, and be aware that the client will most certainly on occasions resort to self-defence.

It will never be possible to avoid all false alarm sources, and the designer can only account to a certain extent for the user and human error, so we should include the client as one source of false alarms:

- the environment;
- the users;
- interconnections;
- the equipment;
- reaction forces.

All of these are sources of false alarms, and systematic attention to each will contribute to an understanding of the problem as a whole and to solutions, which if implemented, would make huge inroads into reducing the false alarm problem. The consequences of not dealing with false alarms include neglect by operating personnel of a true alarm and withdrawal of reaction by the required services. Is it correct to announce an acceptable level when a single alarm creates a problem in some quarter? Perhaps we would be better advised to measure against success or failure. If, as an example, the number of arrests as a direct contribution of the use of intruder alarms was doubled we could see this as success even if the false alarm problem remained the same. If the figure for arrests was doubled and the number of false alarms halved, we could again see this success, but it would still remain that problems exist for those subjected to false alarms. Nevertheless, the fact remains that intruder alarms are vital, and concerted efforts must be made to reduce any incidence of false alarms without any target value. Sadly, no simple statistic can include a measure of crime prevention effectiveness. So although we may hear of the false alarm problem, there is silence regarding the effectiveness of alarm installations. To this end, for operational purposes reaction forces remove true alarms from the equation and concentrate on false alarms themselves. Instead of expressing an incidence of false alarms they compare the number of false alarms against time to establish a false alarm rate per year. Thus, any given installation can be attributed with a target rating of a number of false alarms per year, and if that rate is exceeded the reaction force cover can be withdrawn from the protected premises. It is a necessary approach, and forms part of the ACPO policy.

In so far as the intruder alarm engineer is concerned the client will always assess the quality of workmanship and recognize the need to use a reputable company. For the larger installation the client will wish to see the detailed specification for the system plus a document on the quality of the workmanship written by a security specialist. It is worth remembering that individuals installing security wiring do not of necessity have to be trained electricians, and many are not; however, they should be committed to the installation of systems to the same quality as any fully trained and approved electrician. The intruder alarm engineer is of course governed by standards, and the client must be made aware of this. The installation company must provide and retain secure records of the installation. The documentation referred to is normally referred to as the as built drawings in other electrical industries. These drawings are a record of the exact location, circuit routing and such as established at the conclusion of the commissioning, and they are intended to show any changes made to the original specification or drawings. In order to comply with industry standards they must provide the following information:

- Name and address of the premises and subscriber.
- Type and location of detectors, sensors, processors, power supplies, warning devices, control equipment, signalling device, etc.
- Details of any client-isolated circuit facilities. This refers to sensor or warning devices.
- Details of any entry/exit routes.

A copy of these drawings should be handed to the subscriber, and they should be secured. These should not form part of the normal operation and maintenance manuals, as the details are in excess of those required for an engineer who might use the manual on a day-to-day basis to set and unset the system.

NACOSS NACP 2, *Code of Practice for Customer Communications*, contains guidelines for customer care and to assist firms in reducing customer complaints, unnecessary call-outs and enquiries and false activations of security systems. It also attempts to enhance the integrity and image of security systems by setting out criteria for maintaining effective customer communications. Other NACOSS codes of practice are given in the reference information part of this book (Chapter 10).

NACP 11 is the NACOSS Gold *Supplementary Code of Practice for the Planning, Installation and Maintenance of Intruder Alarms*. The hand-over checklist, completion certificate, historical record, record of operational checks and preventative maintenance report logging data forms are given in Figure 9.9.

Company Name:.. Company Address: ....

....

Telephone No: ....

HAND–OVER CHECKLIST FOR INTRUDER ALARMS SYSTEMS

Customer: ....

Address:...

Job No: .... Date:................................. 20...........

	Tick When Checked	Remarks
1. Check that the installation is in strict accordance with the Specification, complies with BS 4737 and is to a high standard of workmanship.		
2. Check that the subscriber's premises are left in a tidy condition.		
3. Check all detection circuits are clearly indicated.		
4. Log all detection circuit insulation and continuity/resistance measurements.		
5. Check mains connection is permanently connected and NOT by plug and socket.		
6. Check that the supply fuses to the alarm installation have the correct rating.		
7. Check that all batteries are clearly marked with date of installation.		
8. Log the normal DC current loadings of all power supplies.		
9. Remove the mains supply and check that the battery voltage of all equipment is within the specified limits and the system operates normally.		
10. Check that there is adequate standby battery capacity to meet BS 4737.		
11. Check the operation of the audible alarm on system activation and when the hold-off voltage is removed from any self-activating device.		
12. Check remote signalling apparatus ensuring correct transmission and receipt of all conditions.		
13. Check that the engineer only reset facilities are available where applicable.		
14. Check the operation of all tamper detection circuits.		
15. Check every detector for correct operation through to the Control Units.		
16. Check and record the area or volume of coverage of movement/vibration detectors.		
17. Check beam interruption detectors for correct alignment.		
18. Check that system operating procedure is displayed near to Control Unit.		
19. Check the exit/entry route time delay (if used) for correct setting and record times.		
20. Set system. Operate a detector device. Check the resulting alarm condition is correctly signalled.		
21. The subscriber is to be shown the extent of the protection and correct operation of the system.		
22. Check that all documentation in accordance with BS 4737 is correct. (It is recognized that some checks under this heading may be subsequent.)		
23. Record the number of the Certificate issued in accordance with the Council's Rules.		
24. Obtain subscriber's signature acknowledging receipt of the system keys.		
25. Check that all the surplus materials and equipment are cleared from the site.		

GENERAL REMARKS

Engineer:.. Supervisor:..

NACP 11 Att.1

Figure 9.9   *NACP forms: (a) Hand-over checklist; (b) completion certificate; (c) historical record; (d) record of operational checks; (e) preventative maintenance report*

Completion Certificate

Customer: .... Ref:

Address: ....
....
....

System Record Reference: .... Dated:

I confirm that the intruder alarm system has been installed to my satisfaction
and that I received:

1. a demonstration and instructions on how to operate the alarm system.
2. security code number for the alarm system or .... keys.
3. full and comprehensive written operating instructions for the alarm system.
4. a record book (system log book) for the alarm system.

Customer's Signature:

....

Signature on behalf of Installing Firm:

....

Date:

....

NACP 11 Att.2

HISTORICAL RECORD
for
INTRUDER ALARM

Customer ....
Installation Address ....
....
Installer/Maintainer

Record of visits made to the above installation. This record is to be completed by the maintenance engineer at every visit. It is not a customer document.

TYPE OF SIGNALLING: AUDIBLE ONLY/REMOTE SIGNALLING

DATE	DOCKET No	REASON FOR VISIT	SIGNATURE

See next page for a record of operational measurements taken during system commissioning. It is important that this information is retained.

NACP 11 Att.3 (page 1 of 2)

RECORD OF OPERATIONAL CHECKS
(BS 4737 : Section 4.1 : 1987)

Cct/ zone	Location	Resistance (Ohms)		Detector voltage	BATTERY CHARGING VOLTAGE
		Detector	Tamper		

BATTERY CHARGING VOLTAGE

.... VOLTS

POWER SUPPLY CURRENT (NORMAL)

.... mA

POWER SUPPLY CURRENT (IN ALARM)

.... mA

BATTERY CURRENT (MAINS SUPPLY DISCONNECTED)

.... mA

WARNING DEVICES

	INPUT VOLTAGE TO SAB	SAB CHARGING CURRENT CHECKED
1	volts	(Tick)
2	volts	(Tick)

SYSTEM ATTRIBUTES

BELL DELAY	.... m	PERSONAL ATTACK	SILENT/AUDIBLE
BELL DURATION	.... m	RESET*	CUSTOMER/ENGINEER
ENTRY TIME	.... s	LINE FAULT MONITOR	SILENT/AUDIBLE
EXIT TIME	.... s	BATTERY CAPACITY	.... Ah

*See BS 4347, Part 1 : 1986, clause 5.5

SIGNATURES	DATE
ENGINEER	
SUPERVISOR	

NACP 11 Att.3 (page 2 of 2)

Company Name:	Serial Number ....
and Address:	

PREVENTATIVE MAINTENANCE REPORT

Customer: ....

Address: ....

....

Type of Installation: AUDIBLE ONLY/REMOTE SIGNALLING (delete as necessary).
The following checks have been carried out in accordance with BS 4737 as currently in force:

Item	Check	Tick when complete	Remarks
1.	Check the installation, location and siting of all equipment and devices against the specification.		
2.	Check the satisfactory operation of all detection devices including deliberately operated devices.		
3.	Inspect all flexible connections.		
4.	Check mains and standby power supplies including correct charging rates.		
5.	Check control unit for correct operation.		
6.	Check remote signalling equipment (intelligibility to be ascertained).		
7.	Test (where possible) remote signalling equipment to ARC or Police Station.		
8.	Check all audible warning and alarm devices for correct operation.		
9.	Check the alarm system is fully operational.		

The system has been left in full working order apart from the items listed below:

Items not completed at the time of the check must be completed within 21 days of the date shown.

Time arrived ....   Time left ....

Engineer's Signature ....   Date ....

Customer/Customer Representative's Signature ....

SSAIB Form No. 29. Preventative Maintenance Report

Customer:	Invoice to:
Contact & Tel. No.:	Customer No.:
Address of installation:	Type of installation: * Audible only * Remote signalling (* delete as applicable)

The following checks on the intruder/holdup alarm installation have been completed in accordance with PD6662: 2003	Checked	Remarks
1. Check the installation, location and siting of all equipment and devices against the system record, the as-fitted document, and log book.		
2. Check the satisfactory operation of all detection devices including deliberately operated devices.		
3. Inspect all flexible connections and ensure that cable fittings and equipment are secure and adequately protected.		
4. Check mains and standby power supplies including connections and correct charging rates.		
5. Check the remote signalling equipment for correct operation to ARC (if applicable).		
6. Check line fault monitor for correct operation (if applicable).		
7. Check that the ARC keyholder record is correct. If not, record correct details in "Engineers comments".		
8. Check that ARC site detail (name, address etc.) are correct. If not, record details in "Engineers comments".		
9. Check all audible warning and alarm devices for correct operation.		
10. Check the system is fully operational.		
11. Preventative maintenance check completed and system left in full working order.		

ENGINEER'S COMMENTS:

Additional Work Required | Yes/No |

Date: .... Arrival time: .... Departure time: ....

Engineer's Signature: .... Customer/Authorized Signature: ....

Position: ....

FOR OFFICE USE ONLY

Charge Code	Invoice Amount	Invoice No.	Date

It will be noted that although these logging forms do retain certain references to BS 4737 it should be recognized that the actual maintenance check requirements that need to be carried out are quite similar under the BS as those that apply under the ENs. Form No. 29 is the relevant SSAIB document which will be shown to be similar to that applicable to that used within the NACOSS Gold scheme. In fact Technical Specification TS50131-7: 2003 Application Guidelines (i.e. for installers) includes recommendations relating to the maintenance procedures and takes over the original role of part 4 of BS 4737 dealing with maintenance and records. Nevertheless it follows that the requirements of the installation and the essential criteria such as adequate standby battery capacity, signalling and documentation must be in accordance with the EN.

9.5   Test equipment

All alarm installers and service technicians will find it necessary to measure various properties of circuits. Ammeters are used for measuring current, voltmeters for voltage, and ohmmeters for resistance. The most employed meter in the field or at site is the multitester or multimeter, originally called a volt-ohm-milliammeter and abbreviated to VOM. It is a single instrument for measuring voltage, resistance and current, and will be found in analogue and digital form using a precision moving coil needle pointer or LCD display, respectively. Multimeters are available in a huge range of types and prices, and are all reasonably robust although the digital multimeter (Dmm) is more resistant to shock. Meters should always be protected from prolonged exposure to magnetism and be carried with an extra set of leads and must be checked if the meter ever gives faulty or erratic readings. Every time the meter is to be used it should be checked for correct operation. The battery must be charged and the meter leads must not be loose or frayed.

We will describe briefly the stand-alone ammeter, voltmeter and ohmmeter.

Ammeter

Electric current is measured with an ammeter, a name which has been shortened from ampere meter. Because an ampere is usually too large for alarm system circuits, milliammeters or microammeters are used most. All three meters work the same way, differing only in the range of values that they are capable of measuring.

In a series circuit the current is equal at all points, so if an ammeter is placed in series with the rest of the circuit, the same current that flows through the circuit will flow through the meter.

Voltmeters

These give an accurate measurement of voltage levels or difference in potential. Combined meters for measuring voltage and current are called volt-ammeters.

Ohmmeters

These are used to measure resistance directly.

Multimeters

The multimeter allows the measurement of different quantities such as current, voltage and resistance through selection processes on the instrument.

The original and very much time honoured instrument, the analogue pointer-type multimeter, still has its adherents who continue to regard them as the guardians of electrical measurement. Indeed, they continue to have a place in fault finding where the rate at which the pointer travels across the scale (copied but not surpassed by the bargraph indicator) can be more important than its final reading, and also by its ability to 'average out' noise and fluctuations owing to the inertia of the pointer and moving parts. However, digital meters are more accurate and easier to read, factors which are particularly important when trying to assess the condition of sealed rechargeable batteries, where the hydrometer can no longer be used.

10 Reference information

10.1 Wireless intruder alarms

Wireless intruder alarms are also often called radio systems, and work by linking detection devices to radio transmitters which communicate with a central control unit. These require little wiring but are generally more expensive to purchase than their hardwired counterparts. They are economical in terms of installation costs and are becoming increasingly used, so they hold great potential for the future. Wireless signalling systems consist of a combination of radio receivers and transmitters used in conjunction with intruder detectors that are battery powered. A selected system code ensures that the central control unit receiver can only respond to transmitters set to the same code. Accidental or deliberate interference by other transmissions is therefore negated. The effect of cross-talk between adjacent systems and stray radiation is also negated. The system itself is supervised to provide tamper, low-battery and 'jamming' conditions. The system controller provides a specified number of zones and generally features an integral siren. It requires connection to the 240 V AC mains supply but is equipped with a back-up battery source to give power loss protection. Being supervised the sensors report to the controller so that a visual status of the system transmitters is always provided. Movement detectors used in wireless alarms tend to use passive infrared (PIR) detection principles and transmit their alarm signals back to the system controller. These PIR detectors are battery powered, compact and completely self-contained since the transmitter is contained within the detector housing. Such units are fully tamper protected, wall mounted and have detection ranges generally of some 10 or 12 m and employ a recognized walk test facility. Advanced movement detectors also incorporate pairs of normally open and normally closed contacts for optional additional use with magnetic switch contacts, pressure mats or vibration detectors. Transmitters can also be activated by external magnets fitted to doors or windows. In this case the transmitters take the form of a stylish housing and come complete with the reed switch. Inputs are also provided for connection to any detector which has little or no power. In this context we class flush- or surface-mounted magnetic contacts or microswitches plus pressure-sensitive mats. A single-contact transmitter may be connected to more than one set of switches. It usually features an LED, is fully tamperproof and can signal the switch opening. In cases where it is not desirable to mount the transmitter in the vicinity of the door or window for aesthetic reasons, then it can always be hidden under a curtain to protect adjoining doors or windows. Wires may then be run from the detection device to the transmitter only.

Emergency pocket-sized hand-held transmitters are effective anywhere within the signalling range of the system controller whether the alarm is in a day or set condition. These perform the function of portable panic alarms. They are also available in dual format functioning as both a panic alarm and remote controller. Whilst the panic button will activate an output at the receiver at any time (useful for mobile security guards) the remote on/off can cause a trigger capable of setting or unsetting the central control panel.

Wireless interior sirens that plug into an electric socket and receive line carrier signals from the central control unit are available. External weatherproof sirens with or without a strobe light are activated by radio signals from the controller. These are self-activating if tampered with or if any removal from the wall is attempted. Their power is derived from the 240 V AC mains supply, and a rechargeable battery gives power loss protection in the event of mains failure.

Indeed the continued growth of the wireless intruder alarms sector led to many manufacturers designing professional hardwired systems with expanders that enabled the control equipment to be able to connect wire-free zones. Therefore the control equipment could be seen as a unit that was able to use a combination of wired and wireless devices by means of whatever expanders had been added.

Control panels that include facilities for wireless links and accessories are often referred to as hybrid panels. Other system control panels enable complete wire-free security and home management to be implemented. These feature expandable plug-in modules together with two way voice communication via a standard telephone line to allow the user to communicate with a certain amount of preset numbers.

When installing any type of wire-free systems and devices there are certain considerations or tips:

(1) Always check the signal strength and background noise levels from the proposed locations of all detectors before mounting the device at its intended location. If a signal strength measurement is unavailable, test to ensure that the alarm signals are received at the control panel before installing any devices.
(2) Avoid mounting transmitting or receiving devices or aerials under metal lintels at the top of windows and doors etc.
(3) Do not mount transmitters or receivers on metal or on upvc clad metal door or window frames.
(4) In systems where the receiver aerial is mounted vertically, the greatest radio range can usually be obtained by mounting all of the transmitters

so that their aerials are also vertical. In some systems, the aerials are mounted in both orientations so this will not apply.

(5) If the wiring must be run into a transmitter or receiver, try to keep this away from the aerials as any metal object near an aerial will reduce its performance.

(6) Do not alter the length of the aerial or replace the aerial inside any transmitter or receiver with an aerial that is not approved by the system manufacturer. Aerials are designed to a tuned frequency and altering the physical length of it will affect its operating frequency and performance.

(7) Avoid mounting the receiver or the receiver aerial near to electrical distribution equipment, televisions or computer equipment as these can generate background noise. Avoid large metal objects such as water tanks, boilers and radiators as these can prevent radio signals from passing between the transmitter and the control panel.

(8) If a remote aerial is installed ensure that it is in the correct orientation. Do not cut the aerial cable until the signal strength has been checked. Always look at the direction which the cable leaves the aerial and keep running the cable in the same direction and away from the aerial for as far as practically possible. Never clip the aerial cable or any other cable close to the aerial. Do not run the aerial cable next to mains cables.

(9) In properties with very thick walls adequate signal strength cannot be obtained through them. If the receiver or receiver aerial can be mounted high up in a loft space so the signal path is through the floor boards, this can increase the signal strength to an adequate level particularly where the devices are mounted on outside walls.

(10) Look for the ideal position for the detection devices, i.e. a PIR detector will not necessarily need to go at ceiling level as the hiding of cables is no longer a consideration.

Both wired and wire-free interconnections are covered as part of EN50131-1, these interconnections should include intruder, tamper and fault conditions and other communications necessary for correct system operation and monitoring for component substitution.

There are specific equipment manufacturing requirements including:

- Identification of faults
- Control of access to the system functions and indications
- Monitoring of interconnections between system components
- Monitoring for substitution of system components
- Compliance with power supply requirements
- Compliance and identification of all system components to the security grade
- Compliance and identification of all system components to the environmental class

- Supplying specifications for the system installation and maintenance
- Clear documentation

The exact requirements for the identification of faults varies in relation to the security grade, but, in general, system faults, main power supply faults and back-up power supply faults must be identified. This also applies to communicated systems.

In regard to EN50131-1 there is only one area that mentions wire-free and this refers to monitoring the availability of interconnections. The area covered refers to the frequency of polling between the control equipment and all other devices in the system.

The devices need to report to the control panel more frequently as the security grade increases. For portable, deliberately operated devices, however, this requirement has been removed. This allows remote controls and similar devices to be carried by the keyholder away from the protected premises without generating a fault condition.

For battery monitoring according to EN50131-6 this states that the low-voltage signal shall be generated within ten seconds when the storage device voltage falls below the value indicated by the manufacturer.

A means is needed for the recognition of prime power faults and the requirement for a message to be sent to the ARC.

In recognition of the value of wireless systems for the future, the Security Equipment Assessment Panel (SEAP), which is a government interdepartmental body, commissioned the Police Scientific Development Branch (PSDB) to prepare a test specification for wire-free intruder detection equipment. It was intended to prescribe the design requirements, methods of testing and performance levels for medium to high-risk sites. The requirements were to apply to short-range devices (SRD) with power levels up to 500 mW, which facilitate radiofrequency links between components located on the supervised premises. The security grades were to be defined as Grades 3 and 4 to EN50131-1.

The receiver bandwidth refers to the amount of radio spectrum the receiver will 'listen to' and is usually quoted in kHz. The radio frequencies that the alarm system actually operates on is usually quoted in Megahertz (MHz).

- Frequencies used in wire-free alarm systems are specifically allocated for short-range devices and over the years those mostly used have been **173.225 MHz, 418 MHz, 433.92 MHz and 868 MHz.**
- In recent years the most commonly available frequency for short-range low-power radio devices has been **433 MHz.** This has been used for a multitude of equipment including alarms, headphones, radio controlled toys and weather stations etc. The widespread use of this frequency is the main cause of poor radio alarm system operation.

- The new European R&TTE Directive 1995/5/EC brought order by 'reserving' a new narrow band **868.6625 MHz** frequency for alarm systems.

Of course the EN standards now dictate the requirements for radio systems but originally BS 6799, Code of Practice for Wire Free Intruder Alarm Systems contained criteria and dictated the construction, installation and operation of intruder alarm systems in buildings using wire-free links between components such as radio or ultrasonic connections. It divided wireless systems into six classifications. These may still be encountered so it is relevant to know the history of those classifications in the industry as it did bias the European logic.

- *Class 1*. Generally for those installations requiring only one transmitter or detection device or if a latching device is used. The receiver of a Class 1 system is unable to identify the origin of an alarm transmission when more than one transmitter or detection device is used.
- *Class 2*. Used for basic alarm installations it can identify which detector or transmitter created the alarm but it does not provide tamper protection.
- *Class 3*. This tends to meet the requirements of an audible system intended for installation in a residential property to compete with what would be offered by a hardwired version. It is the minimum class of wireless system that an approved installation company should fit.
- *Class 4*. A class which is similar to Class 3 but has additional features that include a supervisory check on each detection devices three times per day. It is only suitable for audible systems.
- *Class 5*. Greater system integrity is offered than that of Class 4 as this class has more frequent supervisory signals. It is generally specified for audible-only applications in residential or small commercial applications.
- *Class 6*. This is similar to Class 5 but has a shorter supervisory fail time. It requires a greater level of system data encryption and narrow bandwidth to be used for the radio receiver. It can be adopted for those installations that require a police response via an alarm receiving centre.

The use of this radiofrequency transmission by wire-free intruder alarm systems is controlled by the Department of Trade and Industry, Radio Regulatory Division, since the systems have to comply with the appropriate regulations. It will be found that many of these supervised radio alarm systems use a microprocessor-based alarm control panel that has similar functions to an already established hardwired counterpart well known within the industry. It also enables someone familiar with a specific hardwired unit to be able to diversify. In fact, some radio panels also have some hard-wired loops and are called generic systems.

Wireless detection devices are not unlike their hardwired equivalents, and the methods of employing and positioning them are identical. Regarding radio components, these are available as a range of units fully compatible with virtually every hardwired control panel on the market. These provide the ultimate in convenience and performance when wiring is not an option. It is therefore possible to purchase radio detectors, transmitters and receivers which are then linked to an existing hardwired control panel and system via the integral contacts in the receiver.

If we look beyond the addition of radio components to an existing hardwired system and towards the fully programmable professional radio security system, then we should consider limitations on the positioning of the main control unit. Where possible it should be installed as near as possible to the centre of the group of detectors/transmitters to achieve an approximately equal signal strength. It should also be at least 1 m from large metal objects such as radiators, and not fixed underground or in a basement. Equally it should not be in close proximity to electrical switch gear, televisions or computers or similar sources of radio interference. The ideal height is some 1.5 m from the floor and at eye level. If the premises are of reinforced concrete or steel construction then a field strength meter must be employed to ensure that the signal strength is adequate. The signal strength necessary will be quoted in the installation manual. The only wiring needed will be for the installation of an unswitched fused spur adjacent to the control panel for the mains connection. An external buzzer or extension speaker if added must be hardwired, otherwise the only wiring will be that to the external sounder/strobe via a stand-alone (SAB) or self-contained (SCB) bell module. This cabling is essentially the same as that we have come to understand in this book. For full flexibility the professional unit also caters for the use of a wireless smoke detector which is active 24 hours a day and initiates a full alarm if activated.

10.2   Environmental protection

All components for outside duty are very much governed by their ability to withstand extremes of climate, changing ambient temperature limits and by their resistance to the weather and ingress of liquids and dusts. If they are to perform as required then they must be designed and mounted with these factors in mind. To some extent the factors also apply indoors, and it is necessary to appreciate the degrees of environmental protection covered as international standards.

The degree of protection is indicated by the letters IP followed by two characteristic numerals (Table 10.1). The first numeral indicates the protection afforded against the ingress of solid foreign bodies and the second the protection against the ingress of liquids. A common rating is IP54,

Table 10.1   *Environmental protection code numbers*

First characteristic numeral	Degree of protection against *solid bodies*	Second characteristic numeral	Degree of protection against *liquids*
0	No protection	0	No protection
1	Protection against ingress of solid bodies larger than 50 mm, e.g. accidental contact by hand	1	Protection against drops of condensation
2	Protection against ingress of medium-sized solid bodies larger than 12 mm, e.g. fingers	2	Protection against drops of liquid falling at an angle up to 150° from the vertical
3	Protection against ingress of solid bodies larger than 2.5 mm	3	Protection against rain falling at an angle up to 60° from the vertical
4	Protection against ingress of solid bodies larger than 1 mm	4	Protection against splashing water from any direction
5	Protection against harmful deposits of dust. Dust may not enter the enclosure in sufficient quantities to interfere with the satisfactory operation of the equipment	5	Protection against jets of water from any direction
6	Complete protection against ingress of dust	6	Protection against water conditions similar to those on the decks of ships
		7	Protected against the effects of immersion in water for a specified pressure and time
		8	Protected against the prolonged effects of immersion in water to a specified pressure

which offers protection against the ingress of dust to the extent that it cannot enter the enclosure in a quantity that could interfere with the satisfactory operation of the equipment. Water splashed against the equipment from any direction will have no harmful effect.

It should be understood that many goods are supplied in housings that are coded to indicate classification when gaskets are correctly seated and the conduit is sealed on wiring. It follows that in a preinstalled state they do not necessarily warrant the applied code.

On occasions one may also see supplementary letters to the IP code. The letter of particular note is W, which draws attention to the fact that the equipment so designated is suitable for use under specific weather conditions. It will be found to have been provided with additional protective features or processes. The specified weather conditions and noted features or processes are agreed between the manufacturer and user and endorsed in data provided by the said manufacturer. For instance, one could bring attention to the circuit board, which may be additionally lacquered for water resistance.

10.3   Multiplication factors

When dealing with electrical circuits the values of current, voltage and power are often expressed in multiples of their units by prefixing a letter to the unit. For example, a current may be written as 10 mA (10 milliamperes) or a voltage as 2 kV (2 kilovolts). Common multiples are shown in Table 10.2.

Table 10.2   *Multiplication factors*

Multiplication factor		Prefix	Symbol
1 000 000 000	10^9	giga	G
1 000 000	10^6	mega	M
1 000	10^3	kilo	k
100	10^2	hecto	h
10	10	deca	da
0.1	10^{-1}	deci	d
0.01	10^{-2}	centi	c
0.001	10^{-3}	milli	m
0.000 001	10^{-6}	micro	μ
0.000 000 001	10^{-9}	nano	n
0.000 000 000 001	10^{-12}	pico	p

10.4   Reference standards, codes of practice and regulations

This section lists the essential documents that have an influence on intruder alarms. It also contains a number of standards that although withdrawn, still have an effect on the industry, because they were time honoured and the standards that will eventually replace them are not as yet complete. Therefore there is a period of parallel running between wrapper documents and the new standards.

The withdrawn BS 4737, the influence of PD6662 and the new suite of:

EN50131 Standards refer. See pages 350–351.

BS EN50132: Alarm systems. CCTV surveillance systems for use in security applications.

BS EN50133: Alarm systems. Access control systems for use in security applications.

BS EN50134: Alarm systems. Social alarm systems.

BS EN50136: Alarm systems. Alarm transmission systems and equipment.

BSIA Form No. 171. Issue 2 May 2005

Guideline for the use of the PD6662: 2004 scheme for the implementation of prEN5013-1: 2004

PD6662 is the foundation document in the UK for the implementation of European Standards for Intruder Alarm Systems – mandatory for all systems installed from 1 October 2005.

This document has been prepared to give guidance on the interpretation of some of the clauses in PD6662 and prEN50131-1: 2004 that may be in need of clarification.

BS 4737 ≠ IEC 60839

BS 4737: Part 1: 1986, Intruder alarm systems in buildings. Specification for installed systems with local audible and/or remote signalling.

BS 4737: Part 2: 1986, Specification for installed systems for deliberate operation.

BS 4737: Part 3: Section 3.0: 1988, Specifications for components. General requirements.

BS 4737: Part 3: Section 3.1: 1977, Requirements for detection devices. Continuous wiring.

BS 4737: Part 3: Section 3.2: 1977, Requirements for detection devices. Foil on glass.

BS 4737: Part 3: Section 3.3: 1977, Requirements for detection devices. Protective switches.

BS 4737: Part 3: Section 3.4: 1978, Specifications for components. Radiowave Doppler detectors.

BS 4737: Part 3: Section 3.5: 1978, Specifications for components. Ultrasonic movement detectors.

BS 4737: Part 3: Section 3.6: 1978, Requirements for detection devices. Acoustic detectors.

BS 4737: Part 3: Section 3.7: 1978, Requirements for detection devices. Passive infrared detectors.

BS 4737: Part 3: Section 3.8: 1978, Requirements for detection devices. Volumetric capacitive detectors.

BS 4737: Part 3: Section 3.9: 1978, Requirements for detection devices. Pressure mats.

BS 4737: Part 3: Section 3.10: 1978, Requirements for detection devices. Vibration detectors.

BS 4737: Part 3: Section 3.11: 1978, Requirements for detection devices. Rigid printed-circuit wiring.

BS 4737: Part 3: Section 3.12: 1978, Specifications for components. Beam interruption detectors.

BS 4737: Part 3: Section 3.13: 1978, Requirements for detection devices. Capacitive proximity detectors.

BS 4737: Part 3: Section 3.14: 1986, Specifications for components. Specification for deliberately-operated devices.

BS 4737: Part 3: Section 3.30: 1986, Specifications for components. Specification for PVC insulated cables for interconnecting wiring.

BS 4737: Part 4: Section 4.1: 1987, Codes of practice. Code of practice for planning and installation.

BS 4737: Part 4: Section 4.2: 1986, Codes of practice. Code of practice for maintenance and records.

BS 4737: Part 4: Section 4.3: 1988, Codes of practice. Code of practice for exterior alarm systems.

BS 4737: Part 5: Section 5.2: 1988, Terms and symbols. Recommendations for symbols for diagrams.

BS 5979: 2000, Code of practice for remote centres receiving signals from security systems.

BS 6707: 1986, Specification for intruder alarm systems for consumer installation.

BS 6799: 1986, Code of practice for wire-free intruder alarm systems.

BS 6800: 1986 (2001), Specification for home and personal security devices.

BS 7042: 1988, Specification for high security intruder alarm systems in buildings.

BS 7671: 1992, Requirements for electrical installations. Issued as the IEE Wiring Regulations 16th ed.

BS 7807: 1995. Code of practice for design, installation and servicing of integrated systems incorporating fire detection and alarm systems and/or other security systems for buildings other than dwellings.

BS 7858: 1996, Screening of personnel employed in a security environment.

BS 7958: 2005. Closed circuit television management and operation code of practice.

BS 8220: 1986/1996, Guide to the security of buildings against crime. Part 1: Dwellings. Part 2: Offices and shops. Part 3: Warehouses.

BS 8418: 2003. Installation and remote monitoring of detector-activated CCTV systems – Code of Practice.

BS EN ISO 9000: 2000, Quality management systems. Fundamentals and vocabulary.

BS EN ISO 9000-1: 1994, Guidelines for selection and use.

BS EN ISO 9001: 2000, Quality systems. Model for quality assurance in design, development, production, installation and servicing.

BS EN ISO 9003: 1994, Quality systems. Model for quality assurance in final inspection and test.

BS EN ISO 9004: 2000, Quality management systems. Guidelines for performance improvements.

DD 243: 2004, Code of practice for the installation and configuration of intruder alarm systems designed to generate confirmed signals.

PAS 020: 1997, Specification for audible personal attack alarms.

British Standard Institute Drafts for Development

BS DD 242. High security.

BS DD 243. Automatic rearming, filtering and substantiating alarm information.

BS DD 244. Wire-free alarms.

BS DD 245. Management of false alarms.

NACOSS Codes of practice

REG (Issue 3) NACOSS Regulations.

NACP 0, Criteria for recognition.

NACP 1, Code for security screening of personnel. Now replaced by BS 7858.

NACP 2, Code for customer communications.

NACP 3 (Issue 2), Code for management of sub-contracting.

NACP 4 (Issue 2), Code on compilation of control manual.

NACP 5, Code for management of customer complaints.

NACP 10 (Issue 2), Code for management of false alarms.

NACP 11, NACOSS Gold Supplementary code for the planning, installation and maintenance of intruder alarms.

NACP 12, Code for wire-free interconnections within intruder alarms.

NACP 13, Code for intruder alarms for high security premises.

NACP 14, Code for intruder alarm systems signalling to alarm receiving centres – superseded by DD 243.

NACP 20, Code for planning, installation and maintenance of closed circuit television systems.

NACP 30, Code for planning, installation and maintenance of access control systems.

Standards and Documents for "EN" Intruder Alarm Systems (October 2006 – As of January 2006)

Document		Number	Called up by PD 6662	Description	BSIA guidelines Form No.
UK Prefix*	Doc type*				
PD	prEN	6662: 2004		Wrapper document specifying documents for the EN50131 scheme in the UK and with some requirements brought over from BS 4737	Form 171 Issue 2
DD CLC/	TS	50131-1: 2004	Yes	*System Requirements*	
DD CLC/	TS	50131-2-2: 2004	Yes[a]	PIR	Form 185
DD CLC/	TS	50131-2-3: 2004	Yes[a]	Microwave	Form 185
DD CLC/	TS	50131-2-4: 2004	Yes[a]	PIR/Microwave	Form 185
DD CLC/	TS	50131-2-5: 2004	Yes[a]	PIR/Ultrasonic	Form 185
DD CLC/	TS	50131-2-6: 2004	Yes[a]	Opening door Contacts (magnetic)	Form 184
DD CLC/	TS	50131-3: 2003	Yes[a]	*Control & Indicating Equipment* (CIE) (i.e. control panels)	Form 179
DD CLC/	TS	50131-5-3: 2005	Yes[b]	*Wirefree* Interconnections (RF) standard	
BS	EN	50131-6: 1997/8	Yes	*Power Supplies*	Form 180
DD CLC/	TS	50131-7: 2003	Yes	*Application Guidelines* (i.e. for installers)	
BS	EN	50136-1-1: 1998	Yes	*Alarm Transmission Systems* (ATS) (i.e. comms to ARC) General	
BS	EN	50136-1-2: 1998	Yes	Using dedicated paths	
BS	EN	50136-1-3: 1998	Yes	Using PSTN	

BS	EN	50136-1-4: 1998	Yes	Using voice communicators	
BS	EN	50136-2-1: 1998	Yes	*Alarm Transmission Equipment* (ATE) General	
BS	EN	50136-2-2: 1998	Yes	Using dedicated paths	
BS	EN	50136-2-3: 1998	Yes	Using PSTN	
BS	EN	50136-2-4: 1998	Yes	Using voice communicators	
DD		243: 2004	Yes	*Alarm Confirmation*	
BS		4737-3-X (excl. −3–30)[c]	Yes	Specifications for *Components* (i.e. parts without an EN)	
			No	"Manuf. Agreement *Additional Comms Req.* between ATE & CIE"	Form 175
			No[d]	"ACPO Requirements For The Reduction of *False Hold-Up Alarms*"	Form 176
			No	Commissioning and Preventative Maintenance (*remote servicing*)	Form 177

* BS = British Standard, CLC = CENELEC, DD = Draft for Development, EN = European Standard, PD = Published Document, pr = Draft, TS = Technical Specification, [a] Included as of 1st October 2006, [b] Included as of January 2006, [c] Also excluding 4737-3-4 and 4737-3-7 as of 1st October 2006, [d] Included in ACPO Security Systems Policy

Index

Printed and bound by CPI Group (UK) Ltd, Croydon, CR0 4YY

03/10/2024

01040432-0017